GOLDEN DREAMS, POISONED STREAMS

How Reckless Mining Pollutes America's Waters, and How We Can Stop It

Introduction by
Stewart L. Udall

Written by
Carlos D. Da Rosa, J.D.
James S. Lyon

Edited by
Philip M. Hocker

with
Thomas J. Aley
Wilgus B. Creath
L. Thomas Galloway, J.D.
Samuel L. Luoma, Ph.D.
Robert Mason, Ph.D.
Glenn C. Miller, Ph.D.
Johnnie Moore, Ph.D.
Karen L. Perry

MINERAL POLICY CENTER
1997

GOLDEN DREAMS, POISONED STREAMS

Da Rosa, Carlos D., 1965 —
Golden Dreams, Poisoned Streams: How Reckless Mining Pollutes America's Waters, and How We Can Stop It / Introduction by Stewart L. Udall; Written by Carlos D. Da Rosa, James S. Lyon; Edited by Philip M. Hocker with Thomas J. Aley...[et al.].
p. cm.
Includes bibliographical references and index.
ISBN 1-889617-01-6 (pbk.)
1. Mineral industries —Environmental aspects —United States.
2. Water —Pollution —United States. I. Lyon, James S., 1953 —
II. Hocker, Philip M., 1944 —. III. Title.
TD428.M56D32 1997
363.739'4 —dc21
97-35901 CIP

Printed in the U.S.A. on recycled paper using soy-based inks.

Front Cover Photographs: Montana Environmental Information Center and Ann Maest; Back Cover Photograph: Greater Yellowstone Coalition.

Table Of Contents

INTRODUCTION

Stewart L. Udall
Chairman, Mineral Policy Center
Secretary of the Interior, 1961-69

G*olden Dreams, Poisoned Streams* is an alarming factual account of the massive damage hardrock mining has wrought on America's water resources.

Dreams of hitting the mother lode drew Americans to the Western frontier to pan for gold more than one hundred years ago. With pick-ax and mule in tow, miners ventured West to search those pristine lands for precious metals. The "forty-niners" of the nineteenth century gold rush have become the garish, glimmering American folklore of the twentieth century. I write about the dirty realities of the California rush in my upcoming book:

"As the pick-and-shovel mining boom petered out, hydraulic mining boomed, and it proved to be the most destructive method of mining the world had ever seen.

"Hydraulic mining moved massive amounts of soil into river systems that drained the Sierra Nevada. For every ounce of gold collected, tons of topsoil and gravel were washed into nearby rivers. When spring floods came, clear streams became clogged with debris, downstream communities were inundated with muck, and fertile farmlands were blanketed with mud and gravel. The town of Marysville, once a pleasant, riverside port for side-wheeler steamboats, illustrates the damage caused by this monstrous machine: When mine wastes filled the channel, the residents were forced to erect protective walls that gradually rose higher than its rooftops. But the depredations of the hydraulic miners were unrelenting, and in 1875 Marysville was buried in muck when a big spring storm overpowered its levees.

"The city dwellers and farmers who suffered severe damage protested, but they got nowhere in a state where gold was king. The gold rush enshrined a value system in California that gave the hydraulic mining industry a right to slice off mountains even if it meant that homes and farms were destroyed, that river navigation was wiped out, and that the ecology of vast tracts of forested uplands was irreparably impaired."

Contrary to popular belief, the American gold rush is still on, and likewise it is no fairy tale. Multi-national mining companies now swarm those same resource-intense lands: they dig, plow, pile, treat and dispose of millions of tons of ore each year — all to fulfill the century-old desire to reap rich harvests from our public lands.

Those golden dreams have created a legacy, a nightmare, of poisoned streams. Mining is destroying America's most precious resource: clean and abundant water. Careless mining, past and present, continues to unleash toxic heavy metals, acidic wastes and poisonous processing chemicals like cyanide into thousands of miles of the Nation's rivers and streams. And mining operations that consume water are draining groundwater aquifers vital to communities' survival.

The result of decades of sloppy mining: a major, insidious threat to the health and ultimate viability of watersheds across the United States. Streams polluted by mining lie orange and lifeless, lakes are filled with toxic, unnatural sediments, and water is undrinkable. Some water wells even run dry.

The environmental movement of the past quarter-century has succeeded on many fronts. Regrettably, that progress has left the hardrock mining industry virtually untouched: I left my post as Secretary of the Interior in 1969. At that time, I wrote that the single most important policy change left for the Nation was the complete overhaul of the 1872 Mining Law — to halt mining's careless damage to our land and water resources. Today, almost thirty years later, Mining Law reform remains unfinished. And, in the meantime, the hardrock mining industry has escaped the needed environmental regulation with which other comparable industries now must comply.

That's why, in 1988, I joined with Phil Hocker to form Mineral Policy Center, now the Nation's leading voice dedicated to cleaning up the environmental problems caused by hardrock mining. *Golden Dreams, Poisoned Streams,* this new report from Mineral Policy Center, describes the massive damage hardrock mining has caused to water resources. The book explains the problem in vivid, factual detail, and it outlines how we can stop the damage. This report is required reading for anyone who cares about the health of our environment.

Dreams and Streams is a wake-up call: America must take action to clean up this problem now, before it becomes insurmountable.

Stewart L. Udall
Santa Fe

Part I
Water Damage from Mining: The Basics

The first half of *Golden Dreams, Poisoned Streams* is our vision of an introductory class, "Hardrock Mining and Water Damage 101." Part I of this book examines, in textbook manner, the issue of water damage from hardrock mining. It explains the nature of water pollution from hardrock mining, how the damage occurs, the scope and costs of the problem, how pollution can be prevented and cleaned up, and what needs to be done to protect our water resources from future degradation. These four chapters provide a cohesive examination of the problems and cures of water pollution from mining.

Part II examines the issue more deeply; it is a collection of technical papers about water damage from mining authored by scientists and regulatory experts.

Note: *Golden Dreams, Poisoned Streams* describes the history, science, environmental problems, and techniques associated with hardrock mining. Hardrock mining encompasses mining for non-fuel materials: metals like gold, silver and copper, minerals like sulfur and phosphate, plus uranium mining. It does not include coal mining. Coal mining also has been a major source of water pollution in the United States. However, unlike hardrock mining, Congress addressed many of the environmental problems from coal mining twenty years ago, when it passed the landmark Surface Mining Control and Reclamation Act.

Hardrock Mining and Water: An Overview

1

"When the ores are washed, the water which has been used poisons the brooks and streams, and either destroys the fish or drives them away."
— GEORGIUS AGRICOLA, WRITING ON MINING IN 1556 A.D.

Earth is the Water Planet — of all the bodies that orbit the Sun, only Earth shines in space with the rich blue of oceans and the sharp white of water vapor clouds. Only on Earth does the persevering force of water steadily shape the continents, form their mountains and valleys, and lay down the sandstones and coalbeds that lie beneath us. And in our solar system only on Earth, the Water Planet, has advanced life developed. We rely on pure, unpolluted water in the most fundamental ways: to carry nourishment to every cell in us, and to rinse away the wastes. To bathe our children.

But for centuries miners, propelled by dreams of golden riches, have poisoned streams, rivers, lakes, and seas in their rush to gouge gold, silver, and countless other minerals from the Earth. The toxic legacy of their carelessness lines the floor of Lake Superior in Michigan, and dangles a poisonous threat over San Francisco Bay. This tragic story is today's news, not just yesterday's history. Recently, in South Carolina, a carelessly-run gold mine spilled ten million gallons of cyanide solution into the nearest river. In 1996, a blowout at a mine in the Philippines flushed millions of tons of sediment onto fishing beds. Today on the island of New Guinea, immense copper mines at work dump so much waste rock and silt into riverbeds that the rivers are forced out of their beds and flood vast areas of rainforest. Check tomorrow's paper for more.

Worst of all, this pollution arises from sheer negligence, not necessity. We know how to mine and refine metals without polluting precious water resources. The technology exists, and it is affordable. All we lack is the resolve to compel the mining industry to use it.

The Threat

Mining pollutes, consumes, and diverts water resources. Across the United States and around the world, vast amounts of water are made unusable every day by mining pollution and wasteful mining practices. According to the U.S. Bureau of Mines, mining has contaminated more than 12,000 miles of rivers and streams and 180,000 acres of lakes in the United States.[1] No one has measured the full scope of mining-caused water pollution worldwide, but we know the list of damaged drainages is very long. And the list is growing rapidly, since few developing countries have the laws or regulatory expertise necessary to prevent mismanagement by multinational corporations in remote locations.

Water pollution from mining devastates communities, wildlife, and the environment. Mining's pollution impact is severe and long-term. It

Toxic heavy metals and acid drainage pour into the Sacramento River watershed from the Iron Mountain Mine, even though mining ceased there in 1963.

PHOTO: KEN W. DAVIS, TOM STACK & ASSOCIATES

can poison water, making it unusable for literally thousands of years. It wipes out fish populations, destroys wildlife habitats, kills watershed vegetation, and contaminates surface and groundwater sources for drinking and other uses.

- A monument to America's mining pollution legacy — Iron Mountain — sits in northern California. Until production was halted in 1963, the Iron Mountain mine site (with both underground and surface mines) produced a wealth of iron, silver, gold, copper, and zinc for nearly one hundred years. It also broke open, perforated, and left behind a mountain of chemically-reactive ore and waste rock that continues today to steadily discharge, or "leach," enormous amounts of acid and heavy metal pollutants into nearby streams and the Sacramento River. The Sacramento deposits these pollutants into San Francisco Bay. The 4,400-acre site accounts for one quarter of all zinc and copper pollutants released each year into U.S. waters from all sources, including all factories and cities.

  Each day, Iron Mountain discharges huge quantities of heavy metals, including 425 pounds of copper, 1,466 pounds of zinc, and 10 pounds of cadmium.[2] Acid waters draining from the site long ago decimated nearby streams, where the acidity in water has been measured as low as <u>minus 3</u> ("-3") on the pH scale[3] — 10 thousand times more acidic than battery acid. Streams downstream from the mine are almost devoid of life.

  Since at least the 1920s, Iron Mountain mine runoff has caused devastating fish kills, poisoning hundreds of thousands of salmon and steelhead trout in the Sacramento River. Despite expensive efforts to reduce and control the pollution at the Iron Mountain (now on the Superfund "National Priority List" for toxic-waste cleanup), enormous amounts of contaminants continue to wash off the site. The problem is so persistent that several experts have reported that at present pollution rates the Iron Mountain site can be expected to leach acid for at least 3,000 years before the pollution source is exhausted.[4]

- Drinking water wells near the Oronogo Duenweg Superfund site, a sprawling complex of lead and zinc mines in Southwestern Missouri, have been contaminated by past mining activities in the region. In 1994, EPA tests revealed that approximately 100 community wells near the site were contaminated with cadmium, lead, manganese, and zinc at toxic levels that significantly exceeded federal safe drinking water standards.[5]

- In Santa Rita, New Mexico, the Phelps Dodge Corporation's Chino Copper mine has long been a menace to valuable water resources. Since being purchased by Phelps Dodge Corporation in 1986, the mine has been plagued by spills, leaks, and discharges of contaminated mine waste material. Much of the pollution has spilled into Whitewater Creek, which runs through a densely populated community. In several incidents in 1987, the mine spilled more than 327,000 gallons of mine wastewater off the site.[6,7] In 1988, another spill occurred, discharging more than 180 million gallons of mine wastewater.[8] More than 90,000 gallons of wastewater spilled in 1989.[9] And in 1992, another 120,000 gallons were spilled.[10]

- Nearly 11,000 fish were killed in 1990 when heavy rains caused a containment pond at the Brewer Gold Mine in Jefferson, South Carolina, to breach, dumping more than 10 million gallons of cyanide-laden water into the Lynches River. Cyanide levels as high as 20 parts per million were detected downstream.[11]

It does not have to be this way. As Agricola recorded more than five hundred years ago, we have known for centuries the damage that mining activities cause to water resources. Today, we also have the knowledge to prevent that damage, and to prevent it at an affordable price. In many cases, current environmental protection technology, if properly applied, will safeguard the purity and quantity of water resources. On the other hand, there are still some situations where, even with the best technology, the pollution risks are too grave. In these cases, mining ventures should not be allowed to go forward.

We cannot leave water protection to chance, or to afterthought by corporate executives whose first commitment is to profits. While cooperative approaches with industry are important, history has shown that the public cannot rely on mining companies alone to manage their mines, their wastewater, their mills and smelters in ways that will protect the purity and quantity of the waters they touch. Ultimate responsibility for protection of this public trust must be in the hands of governmental agencies responsible to the public at large. Unfortunately, American government regulation of mining is woefully inadequate on both the state and federal levels. Abroad, in developing countries, mining regulations are virtually non-existent, or, if in place, often unenforced.

Current U.S. mining regulation is plagued by profound environmental protection gaps, vague requirements, and weak standards. Too often, the enforcement performance of regulatory agencies has been poor and inconsistent. Lured by the promise of jobs and tax revenues, mining regulators (particularly at state and local levels) are often more interested in assisting mining development regardless of the environment, than in regulating the industry to protect the environment.

The problem of water pollution from mining is growing, particularly as the mining industry expands worldwide to meet the steadily increasing mineral demands of the global economy. Today's worldwide free-trade trends permit mining corporations access to new ore deposits in previously-closed developing nations. Not only is mineral production growing worldwide, but new technology enables mining companies to profitably mine low grade mineral deposits that were once unprofitable.

To stem this growing threat of mining pollution, society must improve mining regulations, strengthen environmental standards, and require the mining industry to develop and use better technology to prevent water pollution.

Science Notes

pH pH pH

Acids, Bases, and "pH"

Many of the materials that shape our environment are **acids**, such as lemon juice, vinegar, and the sulfuric "battery acid" in automobile batteries. Another group, the **bases**, includes household detergent, milk of magnesia, and chemical plumbing drain cleaners. Strong acids are poisonous, and they react aggressively to corrode metals and many other materials.

Chemists developed the "pH" system to express the acidity or basicity of materials. The pH scale runs from 1 to 14, although pH values lower than 1 are possible and have been measured. Neutral water has a pH of 7, which is the midpoint of the pH scale. The pH is inversely related to the concentration of hydrogen ions (or acid) in a solution — the more acidic a solution, the higher the concentration of hydrogen ions, and the lower the pH.

Approximate pH Values for Common Materials	
Battery Acid	1
Lemon Juice	2
Acid Rain	3.5-5.0
Pure Water	7.0
Phosphate Detergent	9
Milk of Magnesia	10
Household Ammonia	11
Caustic Soda	13

The pH scale is logarithmic, so a change of only one pH unit is equal to a ten-fold change in acidity. Therefore, a solution with a pH of 2 is ten times more acidic than one with a pH of 3, one hundred times more acidic than one with a pH of 4, and so on. The hydrogen ion concentration is measured in moles of hydrogen per liter of water, or "molar concentration." (A mole is one unit of molecular weight, a standard quantity used in chemistry.) A solution with a hydrogen ion concentration of $1x10^{-2}$ molar, or one one-hundredth, has a pH of 2, because pH is defined as the negative log of the hydrogen ion concentration: The log of $1x10^{-2}$ is -2; the negative of -2 is 2.

The pH of healthy human blood is between 7.35 and 7.45, and small changes outside that range make you very sick. The optimal pH range for the survival and reproduction of most fish is between 6.5 and 8. Streams contaminated with acid mine drainage have a pH of 4.5 or lower and are often devoid of fish and macroinvertebrates.

Source: Lingren, Wesley E., Essentials of Chemistry, Prentice-Hall, 1986, pp. 364-393.

A Wakeup Call

To make an enduring improvement in the way mining is carried out, to ensure that vital water resources are no longer wantonly sacrificed for the sake of golden dreams, we must demand that water protection come before mineral production and miners' profits. Pure water is more precious than gold. Water protection and mineral production can coexist, but not in all circumstances and not without changing how we do business today. Mining companies must be required to demonstrate clearly that their projects will not pollute or unduly waste our most precious earthly resource — water. If those demonstrations cannot be made and maintained, mining should not proceed. Once a mine is in operation water protection must remain the highest goal of the company, even if it means reduced mineral productivity. Adopting this common-sense ethic is the only way we can ensure that the golden dreams of mining do not continue to turn into more nightmares of poisoned streams:

- The DeLamar silver/gold mine in Idaho has repeatedly dumped heavy-metal-laced wastewater into nearby streams, and migratory waterfowl have been poisoned by cyanide from its ponds.[12, 13]
- The Stibnite gold mine, also in Idaho, has leaked cyanide into nearby groundwater and the East Fork of the Salmon River — an important salmon spawning run.[14, 15, 16]
- The Ray Mine in Arizona has polluted nearby groundwater with toxic levels of copper and beryllium. In 1990, rainwater washed more than 324,000 gallons of copper-sulfate-contaminated wastewater from the mine into the Gila River.[17]

We must not allow this careless mismanagement to continue. We need not. This book is a wakeup call for action to protect our planet's precious water resources.

Poisoning Streams

Metals mining requires a series of different actions to uncover and extract mineral deposits in the ground or in stream beds. *Ore* is rock or earthen material that contains veins, beds, or particles of a target mineral, such as gold or copper. In open-pit mining, to get to the ore companies must excavate large quantities of *waste rock* (material not containing the target mineral). After being removed, waste rock, which often contains acid-generating material, heavy metals, and other contaminants, is usually stored on the earth's surface in large open piles. This waste rock and the chemicals used to process the ores represent the source of much of the water pollution caused by mining.

After the waste rock is removed and the ore is extracted, the ore must be processed to separate the target mineral from the valueless portion. Once the minerals are processed and recovered, the remaining rock becomes another form of mining waste, called *tailings*. Mine tailings often contain the same toxic heavy metals and acid-forming minerals that waste rock does. Tailings can also contain chemical agents used in processing the ores, such as cyanide or sulfuric acid. Tailings are usually stored on the earth's surface in piles or containment ponds (though they are sometimes pumped back into the underground space from which the ore was mined). In some cases, mining companies operating overseas in developing countries dump mine tailings directly into rivers or oceans — a practice no longer allowed in the U.S.

Mining waste does not lie inertly in the environment. Exposed to the elements, contaminants in mine waste material can easily leach out into surface and groundwater, causing serious pollution. Sloppy practices at a mine can produce accidents that cause waste rock and tailings to spill or migrate off the mining site, and such releases can result in catastrophic environmental damage. For example, continuous leakage from United Nuclear Corporation's uranium mill tailings impoundment in McKinley County, New Mexico, contaminated three local groundwater aquifers before the site was closed in 1982. Earlier, in 1979, a major dam failure at the tailings impoundment resulted in a massive spill of 90 million gallons of uranium tailings and wastewater into Pipeline Canyon and the Rio Puerco River.[18]

PHOTO: GARY MILBURN, TOM STACK & ASSOCIATES

Trace amounts of heavy metals, such as arsenic or lead, and chemical processing agents like cyanide, can be lethal to fish and wildlife.

The quantity of waste material produced by mining is enormous, particularly in comparison to other waste production sources. Hardrock mining already has produced and put into the U.S. environment approximately 70 billion tons[19] of waste material, enough to cover the entire state of Vermont — rivers, valleys, forests and mountains — with a lifeless layer four feet thick.[20]

And the layer would get thicker and thicker. Each year, the mining industry creates a quantity of waste which is NINE TIMES GREATER than all the municipal solid waste discarded in all the cities in the nation. In 1987, mines in the U.S. dumped 1.7 billion tons of new solid waste onto our lands, while all the waste from all the cities of America totalled 180 million tons[21] (see Chapter 2, Chart).

Much of this waste contains hazardous acid-forming chemicals or toxins. Every year, the hardrock mining industry generates approximately the same amount of hazardous waste as all other U.S. industries combined. According to the EPA, the U.S. hardrock mining industry generated approximately 61 million metric tons of hazardous waste in 1985. In comparison, all other American industries combined generated 64 million metric tons.[22]

Mining pollution occurs primarily in four forms: (See Chapter 3 for more detail.)

Acid Mine Drainage — Acid Mine Drainage, or "AMD," pollution is generated when rock, excavated from an open pit or opened up in an underground mine, which contains sulfide minerals reacts with water and oxygen to create sulfuric acid. Iron pyrite, "fool's gold," (FeS_2) is the most common rock type that reacts to form AMD. The acid will leach from the rock as long as its source rock is exposed to air and water and until the sulfides are leached out — a process that can last hundreds, even thousands, of years. Acid is often carried off the mining site by rainwater or surface drainage and deposited into nearby streams, rivers, lakes, and groundwater. AMD severely degrades water quality, killing aquatic life and making water virtually unusable.

Heavy Metal Contamination — Heavy metal pollution is caused when metals such as arsenic, cobalt, copper, cadmium, chromium, lead, silver, or zinc contained in excavated rock or exposed in an underground mine come in contact with water. As water washes over the rock surface, metals are leached out and carried downstream. Metal leaching is greatly accelerated where acid mine drainage occurs. Heavy metals, even in trace amounts, can be toxic to humans and wildlife. When consumed by living things, the metals can *bioaccumulate,* or build up in living tissue, and be passed through the natural food chain.

Processing Chemicals Pollution — Processing chemicals pollution is caused when chemical agents used by mining companies to separate the target mineral from the ore (such as cyanide, sulfuric acid, or the liquid metal mercury) spill, leak, or leach from the mine site into nearby water bodies. These chemicals can be highly toxic to humans and wildlife and can make water unusable. A teaspoon of 2 percent cyanide solution can be lethal to humans; over two hundred million pounds of cyanide is used in U.S. mining each year. Mercury can no longer be used in the United States to process minerals, but it is still used in some developing countries by independent miners. Though it is forbidden today, mercury contamination of surface and groundwater from old mine sites persists across the United States, from Virginia to Alaska. Mercury contamination is highly toxic to most life, and mercury does bioaccumulate in living tissue.

Erosion and Sedimentation — Sedimentation pollution occurs when unearthed rock and soil are eroded by water or dumped into waterways, and then carried off the mine site into streams, rivers, and lakes. While this mine sediment may or may not contain toxic substances, the sediment itself can cause severe environmental damage by clogging river beds and smothering watershed vegetation, wildlife habitat, and even aquatic life. Sedimentation can also cause severe flooding in river systems.

Diminished Water Supplies

Pollution of water is not the only environmental problem associated with hardrock mining. Mining consumes great amounts of water at all stages of mineral production, and mining activities also can waste vast quantities of surface and groundwater, making water unavailable for other uses. In 1984, the U.S. hardrock mining industry used 2.27 trillion gallons of water for mineral production.[23] This staggering amount is equivalent to 35 years of water consumption by Denver, Colorado (population 952,000).[24]

Mining companies often find that they must import water from sources outside of the mining area to meet their mineral production needs. This will be increasingly true in the future as mining grows and available water sources become scarce. Not surprisingly, in some arid areas of the country, mining is already competing with agriculture and municipalities for limited water supplies.

Ironically, while mining companies need a lot of water for mineral processing, at the same time they often find themselves having to divert substantial amounts of water from entering and flooding surface and underground mines (see Chapter 2). Many mines today burrow so deep into the ground that they operate beneath the *water table* (the level below which the earth is continually saturated with water). They must, therefore, pump and remove groundwater, or *dewater* their underground cavities or open pits to prevent the groundwater from flooding their operations. The volumes of water displaced by even a single mining operation can be enormous. In Nevada, the center of the current U.S. gold mining boom, some mines dewater at a rate of thousands of gallons per minute. Substantial dewatering can drastically lower the water table. This causes wells and springs to dry up, depriving other users of access to water and sometimes causing land *subsidence* (land collapsing due to voids or movement below the surface) as dewatered soil and rock compact.

When groundwater pumping ceases, open pits like this one at the Equity Silver Mine in British Columbia fill with polluted water.

PHOTO: ANN MAEST

- One single U.S. gold mine, Canadian-based Barrick Resources Corporation's Goldstrike Mine in Eureka City, Nevada, pumps as much groundwater out of its vast open mining pit in one year as the entire annual consumption of a city the size of Austin, Texas (population 500,000).[25]

Some of this displaced water may be used in the mining process. At some mines, however, the displaced water is diverted directly into surface streams or stored in collection ponds that encourage evaporation. As a result, mining is dramatically drawing down and wasting vast quantities of groundwater resources that will not be replaced for centuries. This massive displacement of groundwater, particularly in arid areas, will have profound environmental consequences for these regions and for future generations, long after the mines are gone.

A National Problem

For many, mining's "golden dream" still evokes romantic images of rugged individuals panning for flakes of gold in the streams of the Wild West. But today's mining is not a small-scale enterprise of undercapitalized entrepreneurs. Mining today is a capital-intensive big business conducted at a vast scale by multinational mining conglomerates using enormous earth-moving and ore-processing equipment. These companies employ highly technical mining methods. They directly disturb tracts of land that are often many miles square, and the extent of their downstream impacts can be orders of magnitude larger than the area of the mine itself.

These impacts are visited on virtually every state across the country — by mining for gold in South Carolina, phosphate in Florida, kaolin and clay in Georgia, copper in Wisconsin and Michigan, lead in Missouri, platinum in Montana, molybdenum in Colorado, and uranium in New Mexico. In 1995, there were approximately 3,000 operating hardrock mines scattered throughout the United States; of these about 250 were metal mines.[26]

The United States is currently the leading producer of hardrock minerals in the world. Our national output accounts for 13 percent of the value of world's annual non-fuel mineral production.[27] The United States is a top-rank producer of a host of hardrock minerals. Our nation, for example, leads the world in production of molybdenum and phosphate rock, while ranking second in the world in the production of both gold and copper, behind South Africa and Chile, respectively.[28] The United States also ranks third among nations in the production of lead and silver. Other nations exceed the U.S. in production of some metals taken singly, but no other nation matches the diversity of American mining output.

The United States' preeminence in hardrock mineral production has come at a steep price, however.

The Sorry Legacy

While mineral production bolsters our national economy, the bill for mining's past and present environmental impacts goes unpaid. Due to irresponsible mining practices and poor regulation, the mining industry has left behind a sorry legacy of more than 557,000 hardrock abandoned mines in at least 32 states. The cost of cleaning up these sites is projected to lie between $32 billion and $72 billion.[29] While many of these abandoned mines are simple landscape disturbances that pose minimal environmental problems, 16,000 of these sites have considerable surface and groundwater contamination problems that seriously degrade water resources.[30] Over sixty of these abandoned sites pose such severe threats to public health and safety that the EPA has listed them on its Superfund National Priority List (NPL) for cleanup attention.[31]

- In 1986, Canadian-based Galactic Resources corporation opened the Summitville Gold Mine in southern Colorado. The company catego-

Many abandoned hardrock mines desperately need reclamation to stem the flow of mining contamination, but there is still no national program to clean them up.

PHOTO: TOM ZUCCARENO, LIGHTHAWK

rized the mine as a "state-of-the-art" cyanide heap leach gold mine. In reality, the mine was an environmental nightmare. Immediately after gold production began, the protective liner under the massive heap of ore being treated with cyanide solution tore, allowing cyanide to leak into the surface and groundwater. Waste rock and other mining material unearthed and exposed by the mine began generating substantial amounts of acid and heavy metal pollution. The cyanide, acid, and metal pollution flowing from the Summitville mine contaminated more than 17 miles of the Alamosa River, a vital water source for farmers in the region. Galactic declared bankruptcy and abandoned the site in 1992. The State of Colorado, which had provided scant regulation of the mine, immediately requested that the Environmental Protection Agency take over the site under the EPA's Superfund program. As of 1996, EPA has spent over $100 million in efforts to clean up the site. Total cleanup costs for the mine alone are expected to top $120 million.[32] Canadian courts have rebuffed Justice Department efforts to collect these costs to the American public from the now-wealthy speculator who organized Galactic Resources and promoted the Summitville mine venture (see Chapters 6 and 8, Photos).

- Twenty-two cows were killed in July 1995 from drinking arsenic-laden water and sediments from a tailings pile at the abandoned Vossburg Mine near Helena, Montana[33] (see Photo Insert).
- In 1986 two tailings impoundments at the abandoned Cinnabar mine in Idaho leached toxic mercury into Cinnabar Creek, a tributary of the Salmon River. Officials feared that local children wandering onto the site might be poisoned with mercury.[34]

Despite the appalling pollution caused by hardrock abandoned mines, the United States has no national program and no dedicated funding to clean up these sites.

The EPA's Superfund program carries out some Federal cleanup activity at polluting hardrock abandoned mines, but the criteria for getting on Superfund's National Priority List for funding have been narrowly drawn: Only sites posing an imminent hazard to public health in heavily populated areas have made the list. As a result, the overwhelming number of abandoned mines across the land, while seriously polluting water, are not eligible for Superfund cleanups.

Moreover, Superfund is already overwhelmed by the complexities and cleanup costs of existing listed toxic-waste sites. Cleanup demands far outstrip Superfund's existing funds. While Superfund is an important program to clean up mining pollution, its scope will continue to be extremely limited. At a few sites, coal-reclamation funds have been diverted to hardrock cleanup, but this remedy can only be used in special cases. As a result, the bulk of the Nation's abandoned mines will continue to pollute water far into the future.

New Contamination

America's problem of water pollution by mining is not limited to abandoned mines and historic sites. Mines in operation today are enlarging and aggravating the damage. Although the mining industry claims that modern mines employ "state-of-the-art" technology that consistently prevents mining operations from contaminating water, reality belies the claim. While improved technology does exist, it is not consistently used and managed properly by the industry. In its *1985 Report to Congress* on mine waste, the EPA averred not only that past mining activities had created a major waste problem in the U.S., but that some of the very mine waste handling practices contributing to these problems were still being utilized by the mining industry.[35]

PHOTO: CHARLES RAY

At the Stibnite Mine in Idaho, this truck was hauling ammonium nitrate (a blasting agent) when it lost control and crashed into the East Fork of the South Fork of the Salmon River.

Since that report was published, virtually nothing has changed. The industry continues to employ poor mine waste handling methods and no new federal regulations have been enacted to change this situation. As a result, many active mines today continue to spill, leak, or leach pollutants off-site into water resources.

- Since opening in 1988, the Gilt Edge gold and silver mine in the Black Hills of South Dakota has been a serious source of pollution to nearby waters. Shortly after mining began, cyanide leaked into the groundwater and nearby Strawberry and Bear Butte Creeks as a result of torn containment liners, poor mine design, and sloppy management practices.[36] Beginning in late 1992, the mine began generating acid mine drainage which washed off the site during a series of rainstorms and other incidents. Problems continued to mount until in October 1994 and again in May 1995 acid drainage from Gilt Edge's waste piles flowed offsite into Ruby Gulch, making the waters of the creek so acid that its pH was measured as low as 2.1 (extremely acidic).[37,38] In eight short years of operation, pollution from the Gilt Edge Mine has made area streams unable to support a viable fish population, and bottom-dwelling invertebrates have been severely diminished. Despite its abysmal operating record, in 1996 the mine's owner, Brohm Mining Company, sought permits which would allow it to double the mine's size.[39]

- Nevada's current gold mining boom may create a legacy of toxic "pit lakes" which pose a major long-term threat to the state's water resources. The mines are digging immense open-pit cavities to reach large, deep deposits of low-grade gold ore. The lakes will form when the open-pit mines which have dug to levels below the surrounding groundwater table close, and the pumps that kept the pits dry during active operations are turned off. Already, six of Nevada's open-pit mines have filled up with water that is polluted beyond federal drinking water standards and aquatic life standards for heavy metals or acidity.[40] Because of the state's feverish pace of mining, at least 30 new pit lakes will begin to form in the next 20 years. The water quality of these future pit lakes has not been adequately assessed, although it may have serious, long-term consequences for Nevada's groundwater and aquatic life.[41]

Too Little Regulation

The mining industry continues to demonstrate that it is incapable of consistently employing protective environmental standards unless required to by regulation. Unfortunately, a strong regulatory program for the hardrock mining industry does not currently exist in the U.S.

The regulation of coal mining is different. Since 1977, the coal mining industry has been regulated under a comprehensive federal law — the **Surface Mining Control and Reclamation Act of 1977 (SMCRA)**.[42] SMCRA's requirements address virtually all phases and aspects of a coal mine operation — planning and permitting, active mining, reclamation, and post-mining monitoring. SMCRA requires all coal operators to comply with national environmental standards, including a requirement to completely reclaim the land to its original contour and productivity after mining. The law is based on the concept of prevention.

By law, coal mining permits may be granted under SMCRA only if the operator can show clearly how the mine will prevent environmental degradation and comply with all requirements. SMCRA allows for state and federal enforcement. It requires inspectors to conduct a minimum number of mine inspections and to cite operators for all violations of law. SMCRA also requires regulators to deny new permits to mine operators that have outstanding violations and existing environmental problems at other mines. The law also grants citizens essential public participation rights.

While SMCRA has not been enforced consistently by state and federal regulators, it has made an impressive difference in ending much of coal mining's history of gross environmental abuse. Furthermore, the law has not been the financial burden that the coal industry claimed it would when SMCRA was being debated in Congress in the early 1970s. When SMCRA took effect in 1977, U.S. annual coal production was approximately 450 million tons. In 1995, nearly twenty years after the enactment of SMCRA, annual coal production has more than doubled to approximately 1 billion tons.[43]

As required by the antiquated 1872 Mining Law, Interior Secretary Babbitt signs away Arizona land containing $2.9 billion worth of copper for just $1,745.

PHOTO: TAMI A. HEILEMANN, U.S. DEPT. OF THE INTERIOR

No law like SMCRA exists for hardrock mining. Hardrock mining in the United States is currently regulated under an inadequate patchwork of federal and state laws that are riddled with regulatory gaps, vague provisions, and weak standards. As Chapter 8 describes, this “safety net” has more holes than net.

A variety of federal laws addresses particular aspects of hardrock mining. Even taken together, however, these laws are not comprehensive in their approach, and none goes to the core elements of mining as does SMCRA — adequacy of mine location, evaluation of the mining plan, environmental standards for mining operations, on-site water protection practices, reclamation requirements, bonding to ensure that reclamation will be paid for, and so forth. Instead, a patchwork of various federal laws tangentially addresses only a few aspects of hardrock mining. Major regulatory gaps are apparent:

The **Clean Water Act**[44], for instance, only partially addresses off-site surface water discharges. Although the Act set limits on pollutants which can be discharged to surface waters from fixed point sources (like pipes and other outlets), it fails to directly regulate discharges to groundwater — though groundwater contamination is a problem at many mine sites. And the Clean Water Act does not set any operational or reclamation standards for a mine to assure that sites will not continue to pollute water sources when they are abandoned.

The **Comprehensive Environmental Response, Compensation, and Liability Act (CERCLA)**[45], better known as Superfund, affixes liability to responsible parties for severe environmental pollution that threatens public health and safety after the pollution occurs, and CERCLA provides some cleanup funds for these sites. Like the Clean Water Act, it does not address any specific aspect of how mines are opened, operated, or closed, and CERCLA is an after-the-fact program, not a preventative one.

The **Federal Land Policy and Management Act of 1976 (FLPMA)**[46] regulates aspects of natural resource management activities on federal public lands administered by the Bureau of Land Management. While this law applies to mining operations on public lands, the regulations the Interior Department relies on to execute its environmental requirements are woefully inadequate. FLPMA has great potential: it requires that all activities on public lands be conducted so as to prevent "unnecessary or undue degradation" of these lands. But this sweeping environmental protection standard has not been used assertively by the Bureau to regulate the many impacts of large, complicated modern mining operations on federal land.

The **National Environmental Policy Act (NEPA)**[47] is model environmental legislation. However, NEPA only requires environmental *studies* of proposals that are considered to be major federal actions affecting the environment. NEPA does not set any objective standards that a mining company must meet. The law merely requires evaluation of a company's plan and assessment of environmental risk. NEPA evaluations can be very limited. A fully detailed NEPA *Environmental Impact Statement (EIS)* may be required for some mines. In other cases, all that might be required is a much less detailed *Environmental Assessment.* With other mining activities, NEPA may not be applied at all. While NEPA is an important supplement to an actual regulatory framework for mining, it is only a supplement.

The **Resource Conservation and Recovery Act (RCRA)**[48] was intended to regulate all aspects of the Nation's management of solid and hazardous waste. Despite the fact that mining generates almost as much hazardous waste as all other industries combined, Congress in 1980 passed an amendment to RCRA, the "Bevill Amendment," which exempted much mining-related waste from RCRA Subtitle C, the federal program regulating the production, storage, transportation, and disposal of hazardous wastes.

Last and worst, the **1872 Mining Law**[49], which governs the extraction of hardrock minerals on public lands, has no environmental protection requirements at all. In fact, the 1872 law actually subsidizes reckless mining activities on public lands. The Mining Law makes mining the "highest and best use" of public lands regardless of the land's environmental sensitivity or other conflicting uses of the land. The law requires that no royalties be paid to the taxpayers (who own the public lands) for minerals extracted, and it allows mining companies to purchase, or *patent,* public lands from the government for $5.00 per acre or less.

State Regulatory Cheesecloth

Because there is no comprehensive federal regulatory framework for hardrock mining, the primary responsibility for regulating the mining industry has fallen to the states. Like its federal counterpart, state mining regulation is implemented in an inadequate, piecemeal approach. Typically, state mining regulation is plagued by gaps, weak standards, and vague language which is difficult to uniformly interpret.

- Arizona mining regulations are riddled with exemptions such as "Concurrent reclamation of surface disturbances is required *unless this is not practical,* in which case reclamation is to take place within two years of completion of mining or exploration; extendable to 15 years, *if reasonable likelihood exists* that the project will resume."[50]
- Nevada's mining law contains no specific standards for reclamation, instead allowing reclamation to be determined on a mine-by-mine basis. The State Department of Environmental Protection is granted the right to approve "any *appropriate* method of reclamation... consistent with the [regulations]" proposed by the mine operator.[51]
- Colorado's reclamation requirements include vague language such as, "regrading shall be *appropriate* to final land use," disturbances to the hydrology and to surface and groundwater *"shall be minimized,"* and reclamation shall be completed "with all *reasonable* diligence... concurrently... to *the extent practicable, taking into consideration* the mine plan, mine safety, *economics, available equipment and material,* and other site-specific conditions relevant and unique to the affected land and to the post-mining land use."[52]

The problem of weak laws is compounded by state regulatory agencies that are frequently understaffed, under-funded, and lack the technical expertise necessary to evaluate and monitor mining activities adequately. Worse yet, some state mining regulatory agencies have an appalling history of being too cozy with the mining industry; they are often more interested in appeasing mining companies than regulating them.

While some states have recently worked to improve their mining regulations, no state can be said to be an environmental model. In fact, while some states move forward, mining industry pressure is forcing other states to move in the opposite direction. During its 1995 session, the Montana State Legislature, under strong industry pressure, severely weakened its mining program, particularly with respect to water quality protection.[53] In 1996, in an effort to overturn this assault on their state water protection laws, citizens organized a ballot initiative (I-122)[54] to strengthen water quality regulations, particularly for mining. However, this initiative was defeated at the polls after the mining industry weighed in and spent more than $3 million in a campaign against it.

As a result of poor mining practices and lax regulation, mining pollution is a national problem in search of a solution. And it is a growing problem. Since the cumulative amount of mining, and mining waste, will continue to grow across the country so will incidents of water pollution from mining. Many states already produce more than 10 million tons of mining waste each year, including Alaska, Arizona, California, Florida, Idaho, Michigan, Minnesota, Missouri, Montana, Nevada, New Mexico, North Carolina, South Dakota, Tennessee, Utah, and Wyoming. Others, including Georgia, Indiana, Iowa, Kansas, Kentucky, Mississippi, New York, Ohio, Oklahoma, Pennsylvania, South Carolina, Texas, Vermont, Virginia, and Washington, produce nearly one million tons of mining waste per year.[55] These amounts are projected to grow dramatically (see Photo Insert and Chapter 2, Maps).

A Global Problem

As Chapter 9 shows, hardrock mining is a worldwide industry and a worldwide environmental problem. The value of worldwide hardrock mining production in 1987 was approximately $140 billion. By 2010, global mining production is projected to reach $200 billion.[56] This growth will bring a staggering increase in the amount of mining waste that is generated and put into the environment.

Highly industrialized countries like Canada, Australia, and South Africa rank with the United States as world leaders in the production of hardrock minerals. These countries also share with the United States a legacy of water pollution from mining.

- On 22 February 1994, fourteen people were killed when a tailings dam collapsed during a rainstorm at the Harmony Gold Mine in the Orange Free State of South Africa. The dam failure sent a six-foot wall of tailings and mine sediment crashing down into a mine worker housing complex.[57]

- In Australia, tailings dams at Kalgoorlie Consolidated Gold Mines' Super Pit (the largest gold mine in Australia) are leaking hundreds of thousands of tons of water contaminated with cyanide and heavy metals into the environment each year.[58]

These industrial countries have done no better than the United States has in preventing mining pollution from contaminating water. And the impact of hardrock mining on developing countries may be even more severe than it has been on the industrialized nations since an increasing portion of the world's mineral production growth is expected to come from developing countries, yet few developing countries have a modern regulatory structure for controlling mining pollution.

These legal inadequacies are exacerbated by the fact that developing country governments seldom have the technical expertise necessary to adequately evaluate and monitor the design, operation, and reclamation of modern, complex mining operations. Instead, these governments often look to the mining companies themselves for advice and monitoring — which invites conflicts of interest.

This 1995 dam break at the Omai Mine in Guyana spilled 2.6 billion liters of cyanide-laced wastewater into the country's main river.

PHOTO: PHILIP M. HOCKER/ MINERAL POLICY CENTER

With little regulatory structure in place, it is not unusual for developing countries to base their environmental protection efforts merely on business contracts with mining companies. These contracts often spell out the financial arrangements and general scope of the proposed mining operations. The contracts may or may not specify what the companies will do to protect the environment. When an individual contract does address environmental cares, it may only vaguely state a company's intentions in broad platitudes, such as "the company will take necessary steps to protect the land and water from adverse impacts from the mine." Without set regulatory standards such vague contract provisions will provide little protection after a poorly designed or operated mine begins to contaminate water.

Sometimes, mining companies do agree to prepare environmental analyses or environmental impact statements. But again, without any regulatory requirements, these environmental assessments are rarely adequate. Due to the absence of international environmental standards, mining companies sometimes now attempt to deflect criticism by claiming that their operations will meet "North American standards," but that is a meaningless term with no clear definition or basis in law.

Because regulatory officials often develop close personal relationships with the miners they are supposed to oversee, it is essential that mine operations and permits be made public. However, developing countries rarely ensure that citizens will be given access to basic information about mining projects or the right to review and participate in government activities related to the mines, such as permitting, monitoring, or environmental protection planning. When things go wrong, it is often nearby communities that bear the burden of polluted or diminished water sources — communities whose citizens have been effectively shut out of the mining and regulatory process.

The problem of maintaining high regulatory standards is made more difficult in developing countries by the high level of economic dependency that some nations have upon foreign exchange earnings from mining. From 1988 to 1990, hardrock minerals accounted for over 99 percent of the total value of Zambia's exports, for example. In Zaire, mineral exports constituted 80percent of total export value. And in Papua New Guinea, Chile, and Guyana, mineral exports were recently 72 percent, 57 percent, and 41 percent, respectively, of total national export values.[59]

In some cases, the governments of developing countries become business partners in mining ventures. This sets up a confusing and conflicting role for the government which is now mine investor and mine regulator. For example, the Guyanese government is a 5 percent owner in the Omai gold mine, South America's second largest. In August 1995, a major environmental disaster occurred at the mine when a dam that held a lake full of tailings failed, spilling more than 860 million gallons of cyanide-laced mine waste into the Essequibo River, the nation's primary waterway. The spill killed fish, produced panic in Guyana's seafood export market, and caused major problems for many residents who depend on the river for drinking water, fishing, irrigation, and transportation. Tailings sediments were reported as far as 50 miles down river.[60]

Guyana had no national environmental protection statute nor any adequate mining regulations in place. The mining operation was governed by a vague contract between the government and the mine operator, Canadian-based Cambior Inc. Cambior claimed that the tailings dam was designed and constructed to meet "North American standards." The government had relied on the company for most of its information and technical expertise about the mine. The nation's heavy economic dependency upon the mine (the Omai mine's output alone made up 20 percent of Guyana's gross national product[61]) also created an incentive for the government to keep the mine operating despite its environmental problems.

Some crude mining practices which are no longer permitted in industrialized countries continue to be used in developing nations due to weak or non-existent mining regulation.

These practices include the dumping of raw tailings directly into rivers, lakes, and the ocean. Additionally, dangerous mineral processing techniques, such as mercury amalgamation (see Chapter 3, Chemical Processing Agents), continue to be used by independent small-scale miners, as in the Amazon region of South America.

A Call to Action

Mining already has contaminated over 12,000 miles of surface rivers and streams in the U.S. Unknown is the volume of groundwater that has been polluted around the world by mining operations. Between 1961 and 1975 more than 10 million fish were killed in the United States alone by mining pollution.[62] And despite the availability of new environmental technology, the problem of water pollution around the world continues to grow today.

It doesn't have to be this way. We can have a successful mining industry while we diligently protect the quality and quantity of our water sources. Chapter 4 discusses known methods to prevent water pollution from mining that, if employed consistently throughout the mining industry, would eliminate hardrock mining's gross environmental abuses.

To ensure that these methods are employed, we need a comprehensive regulatory program for mining — in the United States and in the form of worldwide standards — to prevent and control mining pollution. We also need to establish a *Hardrock Abandoned Mine Reclamation Program*, adequately funded, to clean up America's legacy of polluting abandoned mines.

All of these tools will only be sufficient if we adopt an ethic that places a higher value on protecting the purity and quality of our water than on gouging the maximum corporate profit from a mine.

Unless we put all these pieces into place, the earth's precious water supplies, upon which we all depend, will continue to be polluted, consumed, and diverted by mining. The time to act is now.

♦ ♦ ♦

Notes

1. Robert L.P. Kleinmann, *Acid Mine Drainage,* ENGINEERING AND MINING JOURNAL, July 1989, p.161.

2. U.S. ENVIRONMENTAL PROTECTION AGENCY, OFFICE OF SOLID WASTE, IRON MOUNTAIN SITE, MINING WASTE NPL SITE SUMMARY REPORT, p.5 (1991).

3. D. KIRK NORDSTROM ET AL., MEASUREMENT OF NEGATIVE pH AND HIGH METAL CONCENTRATIONS OF EXTREMELY ACIDIC MINE WATERS FROM IRON MOUNTAIN, CALIFORNIA (ABSTRACT, ANNUAL MEETING OF GEOLOGIC SOCIETY OF AMERICA, SAN DIEGO, CALIFORNIA, 1991).

4. D. Kirk Nordstrom et al., *The Production and Seasonal Variability of Acid Mine Drainage From Iron Mountain, California: A Superfund Site Undergoing Rehabilitation, in* ACID MINE DRAINAGE: DESIGNING FOR CLOSURE, p.18 (21 June 1991).

5. Telephone communication with Mark Doolan, Regional Project Manager, Oronogo Duenweg Superfund site (4 November 1995).

6. U.S. Environmental Protection Agency — Region 6, Administrative Complaint, Findings of Violation, Notice of Proposed Assessment of a Civil Penalty, and Notice of Opportunity to Request a Hearing Thereon, Docket No. VI-90-1616, (3 November 1989).

7. Dorsey Griffith, *Mine's Deal with State Avoids Fine,* SANTA FE NEW MEXICAN, 30 July 1989.

8. ROBERT H. REEVES, U.S. ENVIRONMENTAL PROTECTION AGENCY, NPDES COMPLIANCE INSPECTION REPORT (on Chino Copper mine) (15 December 1988).

9. New Mexico Water Quality Control Commission and Environmental Improvement Division of the New Mexico Health and Environment Department vs. Phelps Dodge Chino, Inc. dba Chino Mines Company, Complaint for Civil Penalties, pp.7-8 (16 January 1990).

10. Letter from D.P. Milovich, Chino Mines Company, to Steven J. Cary, New Mexico Environment Department (10 August 1992).

11. South Carolina Wildlife & Marine Resources Department, *Wildlife Commission Approves Report on Cyanide Fish Kill,* NEWS RELEASE #90-320, 26 November 1990 p.2.

12. *DeLamar Mine Indicted in Waterfowl Deaths on Cyanide Pond,* IDAHO STATESMAN, 18 May 1991.

13. In the Matter of: NERCO DeLamar Company (Environmental Protection Agency 1991) (Compliance Order and Request for Information No. 1090-11-23-308/309).

14. Letter from Idaho Department of Health and Welfare to Hans Gaertsema, Mine Manager, Pioneer Metals Corp. (13 December 1989).

NOTES

15. Letter from James R. Tomkins, Mine Manager, Pioneer Metals Corp. to Veto J. Lasalle, Forest Supervisor, U.S. Forest Service, Payette National Forest (27 July 1990).

16. Letter from Gregory Kellogg, Acting Chief, Water Permits and Compliance, U.S. Environmental Protection Agency, to Don Bork, General Manager, Stibnite Mine, Inc. (23 October 1992).

17. *ASARCO Mine Spills Chemical Into River, Tribe Fears for Crops,* THE ARIZONA REPUBLIC, 16 August 1990, p.A1.

18. U.S. ENVIRONMENTAL PROTECTION AGENCY, UNC SITE UPDATE (May, 1990).

19. Seventy billion ton figure is an update of 55 billion tons of accumulated hardrock mining waste in the United States, as reported in 1985 by the U.S. Environmental Protection Agency. (U.S. ENVIRONMENTAL PROTECTION AGENCY, REPORT TO CONGRESS..., p.ES-17; see note 22, below). Mineral Policy Center estimates that between 1 and 2 billion tons of hardrock mining waste have been generated yearly in the United Stated since 1985, based on U.S. Bureau of Mines and U.S. Geological Survay statistics.

20. Comparisons based on 70 billion tons of mining waste with an average density of 1.8 short tons/cubic yard. (A short ton equals 2,000 lbs.) Density figure is the average density of gold ore with a 20percent porosity. Density figures were provided by Jonathan G. Price, Director and State Geologist, Nevada Bureau of Mines and Geology (26 September 1995).

21. OFFICE OF TECHNOLOGY ASSESSMENT, MANAGING INDUSTRIAL SOLID WASTES FROM MANUFACTURING, MINING, OIL AND GAS PRODUCTION, AND UTILITY COAL COMBUSTION, p.10 (February, 1992).

22. U.S. ENVIRONMENTAL PROTECTION AGENCY, REPORT TO CONGRESS, WASTES FROM THE EXTRACTION AND BENEFICIATION OF METALLIC ORES, PHOSPHATE ROCK, ASBESTOS, OVERBURDEN FROM URANIUM MINING, AND OIL SHALE, p.4-49 (31 December 1985) [hereinafter REPORT TO CONGRESS].

23. U.S. BUREAU OF MINES, UNITED STATES DEPARTMENT OF THE INTERIOR, WATER USE IN THE DOMESTIC NONFUEL MINERALS INDUSTRY, IC 9196, p.35 (1988).

24. Figures based on 65.3 billion gallons of water used by metropolitan Denver in 1995. DENVER WATER, STATISTICAL SUMMARY 1986-1995, p. 37 (1996).

25. U.S. ENVIRONMENTAL PROTECTION AGENCY, FEDERAL REPORTING DATA SYSTEM, WATER SYSTEM INVENTORY (1994).

26. MINE SAFETY AND HEALTH ADMINISTRATION, DEPARTMENT OF LABOR, EMPLOYMENT/OPERATIONS IN MINING INDUSTRIES (28 July 1995).

27. Fact Sheet (undated), *citing* UNITED NATIONS COMMISSION ON TRADE AND DEVELOPMENT SECRETARIAT (1990).

28. METALS & MINERALS ANNUAL REVIEW (1996).

NOTES

29. Mineral Policy Center (MPC) estimates. See MPC methodology in MINERAL POLICY CENTER, BURDEN OF GILT, pp.28-31 (June, 1993).

30. *Id.*

31. U.S. ENVIRONMENTAL PROTECTION AGENCY, MINING SITES ON THE NPL (August, 1995).

32. Berny Morson, EPA, *Recalling Summitville, Gets Tough,* ROCKY MOUNTAIN NEWS, 29 January 1996, p.5A.

33. Mark Goldstein, *Arsenic Killed Cows,* HELENA INDEPENDENT RECORD, 17 August 1995, p.1A.

34. PIONEER TECHNICAL SERVICES, FINAL PRELIMINARY ASSESSMENT FOR THE CINNABAR MINE SITE, STIBNITE, IDAHO (21 May 1992).

35. REPORT TO CONGRESS, *supra* note 22.

36. Barbara Ordahl, *Contamination Traces Found at Brohm Well,* RAPID CITY JOURNAL, 30 December 1989.

37. George Ledbetter, *Acid Drainage at Brohm Mine,* LAWRENCE COUNTY CENTENNIAL, 21 October 1994, p.A1.

38. George Ledbetter, *Flooding Revives Acid Problems at Mines,* LAWRENCE COUNTY CENTENNIAL, 13 May 1995.

39. George Ledbetter, *Expansion Would Double the Size of Brohm Mine,* LAWRENCE COUNTY CENTENNIAL, 7 June 1995.

40. Glenn C. Miller et al., *Understanding the Water Quality of Pit Lakes,* ENVIRONMENTAL SCIENCE AND TECHNOLOGY, March 1996, pp.118-123.

41. *Id.*

42. 30 U.S.C.A. §§ 1201 et seq.

43. ENERGY INFORMATION ADMINISTRATION, U.S. DEPARTMENT OF ENERGY, COAL INDUSTRY ANNUAL 1995, p.5 (October, 1996).

44. 33 U.S.C.A. §§ 1251-1387.

45. 42 U.S.C.A. §§ 9601-9675.

46. 43 U.S.C.A. § 1732(b).

47. 42 U.S.C.A. §§ 4321-4370d.

48. 42 U.S.C.A. §§ 6901-6992k.

49. 30 U.S.C.A. §§ 21 et seq.

50. Arizona Comp. Admin. R. & Regs. R11-2-204, emphasis added.

51. Nevada Admin Code § 519A.285 (1991), emphasis added.

NOTES

52. Colo. Rev. Stat. § 34-42-116(7) (Supp. 1994), emphasis added.

53. Senate Bills S330 and S331, passed by the Montana Legislature in 1995.

54. Montana Clean Water and Public Health Protection Act of 1996 (1996).

55. U.S. BUREAU OF MINES, MINERALS YEARBOOK 1990, p.65 (1991).

56. MAMADOU BARRY, ECONOMIC SIGNIFICANCE OF MINING AND MINERAL MARKET TRENDS, p.2 (1 June 1994) (Prepared for International Conference on Development, Environment, and Mining, Washington, D.C., 1 June 1994).

57. *Harmony: Search Continues,* MINING JOURNAL, 4 March 1994, p.154 .

58. Geraldine Capp, *Huge Mine Pollution Claim,* WEST AUSTRALIAN, 3 December 1996.

59. Fact Sheet (undated), *citing* UNITED NATIONS COMMISSION ON TRADE AND DEVELOPMENT (1990).

60. Desiree Kissoon Jodah, *Courting Disaster in Guyana,* MULTINATIONAL MONITOR, November 1995, pp.9-12.

61. *Guyana,* MINING JOURNAL (Advertisement Supplement), 6 September 1996, p.8.

62. U.S. ENVIRONMENTAL PROTECTION AGENCY, FISH KILLS CAUSED BY POLLUTION, FIFTEEN YEAR SUMMARY 1961-1975, p.8 (April, 1978).

♦ ♦ ♦

A Closer Look at Mining Practices

2

The deep open pit of BHP Copper's Pinto Valley copper mine near Miami, Arizona, constantly buzzes with the sounds of mining. Blastholes are steadily bored into the ore and waste rock; four times a week, the holes are loaded with powerful explosives and detonated. The mine's fleet of earthmoving equipment — 5 large electric shovels, one front-end loader, and 19 haul trucks — works 24 hours a day, seven days a week, to load and haul the blasted material from the mine pit to the copper mill for processing, or to waste rock piles.[1]

Each of the mine's electric shovels is equipped with a bucket that, in a single scoop, can gather 43 tons of broken ore and waste rock — the weight of 33 Honda pickup trucks.[2] *At full speed, it takes the shovel less than five minutes to load a house-sized haul truck, which can carry up to 190 tons of material.*

Pinto Valley's earthmoving fleet moves over 150,000 tons of waste rock and copper ore every day, 60 million tons of material per year. Since the Pinto Valley mine's copper ore contains less than 0.4 percent copper,[3] *the vast majority of this material is eventually discarded as waste. In one year, this single mine generates a mass of solid waste greater than one-quarter of that generated by all the cities and towns in the United States combined.*[4] *The mine's wastes, steadily accumulated over decades, now sprawl over an area as great as 950 football fields.*[5]

These wastes are not simply mounds on the landscape, they are pollution accidents waiting to happen. In January 1993, disaster struck the Pinto Valley mine when runoff from heavy rains rushed through the mine's waste impoundments and into nearby Pinto Creek. The polluted stormwater from the mine fouled Pinto Creek with hundreds of tons of metal-contaminated waste.[6]

Hardrock mining began way back in American history, and so did mining's destructive effect on water. It continues today. The wastes from mining operations of the past that litter the American landscape will continue to pollute water nationwide for hundreds, even thousands, of years into the future.

Over its lifetime, the American hardrock mining industry has generated approximately 70 billion tons of mining waste — material that now lies in unreclaimed and unprotected dumps and ponds across the country. More waste is being added every year. Too much of this waste — which contains acid-forming rocks, heavy metals, and leftover mineral processing chemicals such as cyanide — gets into the nation's rivers, lakes, streams, and groundwater, leading to severe water pollution.

However, mining's impact on water resources is not only the result of yesterday's irresponsible mining practices. Today's mining operations also pose a grave long-term threat. Hardrock mines currently generate over 2 billion tons of waste per year. Many of the same damaging mining methods that were used in the past are still employed by the mining industry today. And the much-greater scale of today's mines, like the Pinto Valley mine described above, poses a new-scale threat to water resources by generating unprecedentedly huge volumes of waste.[7]

Mining waste does not disappear after the miners have no more use for it. Much of it is simply left behind, exposed to the natural elements on the earth's surface. That exposure too often means that pollutants in the waste spread through the environment, contaminating rivers, lakes, and streams. In addition to surface impacts, mining waste also contaminates underground water resources through the leaching of contaminants contained in mining waste.

To understand how water pollution is caused by mining, and how it can be stopped, it is important to understand how hardrock mining is carried out. This chapter is not a technical treatise on mining engineering; there are many of those in print. Rather, it is intended to give a brief overview of some common mining methods, so the water-pollution damage the industry causes can be understood in context.

The Basic Process of Mining

Hardrock mining involves uncovering and extracting mineral deposits in solid ore or eroded deposits in streambeds. These deposits are extracted through either surface mining (extracting ore by digging an open pit), underground mining (digging underground tunnels or shafts), or placer mining (recovering minerals deposited in streambed sediments). With surface and underground mines, mine operators usually must remove and set aside waste rock (material not containing enough of the target mineral to pay for the cost of further processing) in order to gain access to the ore containing the target mineral. The amount of waste rock that must be removed at a modern mine varies, but is often enormous.

With the waste rock out of the way, the ore is removed by blasting it free from the earth (or using other excavating methods or equipment), and transported by truck or conveyer system to a processing area. The ore is then processed to separate the target product from the rest of the ore. In the ore itself, the portion that is salable can vary from almost one-half, as in high-grade iron oxide deposits, to tiny fractions like the low-grade gold deposits at the Zortman-Landusky Mine in Montana, which makes a profit on 0.017 ounces of gold per one ton of ore. The remaining 31,999.98 ounces, which does not bear gold, becomes mining waste.

Mineral processing can be done in various ways depending on the specific target mineral, the quality and grade of the ore, and the experience, equipment, and preferences of the mining operator. Some typical steps include crushing the ore, treating it with chemicals, removing impurities with extreme heat ("smelting"), and refining the mineral electrically to a high level of purity. Each of these methods is discussed in greater detail below.

Each step of the mining process generates waste. In addition to the rock and earth that must be stripped away to get at the ore, waste products include mine tailings, rock and sludge left from mineral processing activities (to separate and refine the mineral from the ore), and chemical processing agents used in the mineral processing stage.

The pollution problems of mining arise both from the specific chemistry of the wastes, and from their immense and growing volume. Today, the pick and shovel of the historic prospector have been replaced by machines that can gather more than 40 tons of material in one scoop and in five minutes can load a 200-ton capacity truck. By moving rock and ore more quickly and efficiently than ever before, today's powerful new equipment makes it possible to mine lower-grade ores than could be used in the past, and still make a profit. This trend is accelerated by the fact that many of the richest, most accessible mineral deposits in the United States have already been mined-out, and the grades of most ores mined today are of necessity lower than the ore grades mined in the past.

Four centuries ago, for example, most copper ores that were mined worldwide typically contained 8 percent copper by weight; today, most copper ores mined average less than 1 percent copper.[8] This does not mean that we use less copper — in fact, around the world more copper is produced each year now than ever before.

Consequently, ore is mined now at a faster rate than ever, and more waste is generated today for every ton of ore that is mined. Even though some of the worst water-polluting practices of the past are now outlawed, the spiraling volume of mine waste threatens to overwhelm the feeble safeguards that recently have been put in place.

American Mining Before the California Rush

The history of mining in America is a long one, covering many minerals. As early as 4,000 B.C., Native Americans mined copper deposits on Michigan's Keweenaw Peninsula and on Isle Royale along the shores of Lake Superior.[9] Early English explorers found iron ore deposits in North Carolina in the late 1500s, and the first known colonial iron works was set up by English settlers near Jamestown, Virginia, around 1620. That crude facility was abandoned after an Indian attack, but small iron works near local ore deposits spread throughout the colonies to supply metal for tools and other implements.[10] The American iron industry grew quickly after Independence. However, it was not until the late 1840s, when huge iron ore deposits were discovered in Upper Michigan and Minnesota, that the United States emerged as a major producer of cast iron and steel. These mineral resources subsequently played a significant role in rapid industrialization and expansion of the United States' infrastructure.

Lead and copper mining also appeared in the colonial period. In the late 17th century, French voyageurs, crossing the continent's midsection in search of furs, began exploiting lead deposits in the Upper Mississippi Valley as raw material for bullets. Later, lead mines were established in present-day Missouri, which is still the center of U.S. lead mining.[11] Small-scale copper mining began in New England in the early 18th century. The earliest colonial copper mine, with an 80-foot underground shaft, was established near East Granby, Connecticut, in 1709.[12] As with iron ore, significant copper mining did not occur until the 1840s, when major deposits of the metal were found in Upper Michigan's Keweenaw Peninsula and in southeastern Tennessee.

Gold mining began in the United States after the discovery of a large nugget in a North Carolina streambed in 1799. An Appalachian gold rush followed, which reached as far south as Alabama, although this rush was much smaller than the later migration to the West. Appalachian gold was mined by panning and dredging of stream gravels (placer mining). There were also a few underground mines — in North Carolina, Virginia, and Georgia — that used slave labor.[13] At this time, gold was typically concentrated using the mercury amalgamation process (see Chapter 3, Chemical Processing Agents). Today, one can still find Appalachian streams that are contaminated with mercury used in this process. After a few decades, interest in gold mining declined in the eastern United States, as the Western goldfields drew the attention of most miners. Gold mining returned to the South Carolina Piedmont in the 1980s and 1990s, however, and renewed gold exploration is underway today in parts of Appalachia.

Eastern mining for calcium phosphate, which is used to produce fertilizer for agriculture, began later. The industry made its start in the early 1880s in central Florida, when phosphate pebbles were found in the Peace River. Later, extensive layered deposits of phosphate were found below the land surface in central Florida. A phosphate rush followed, as scores of companies began to exploit the state's rich resources. Soon, a local fertilizer industry sprang up, supplied by nearby phosphate mines. While the number of mining companies soon dwindled, mining production surged, as the industry increasingly began to employ large, mechanized earthmoving equipment to extract phosphate ore.[14] Today, most phosphate mining takes place in Florida and North Carolina. Smaller quantities of phosphate rock are also mined in Idaho, Montana, and Tennessee.

Mining Moves West

The discovery of gold at Sutter's Mill, California, in 1848 greatly intensified the pace of hardrock mining in the United States. Enticed by the find, thousands of prospectors flocked to the West to seek their fortune, setting off one of the largest human migrations in American history. Most prospectors didn't find their pots of gold, but this expanded mineral exploration led to discoveries of valuable deposits of silver, and later copper and lead. These finds were the bedrock of a growing hardrock mining industry. Consequently, the generation of mining waste grew substantially in the West. The waste was tossed into the environment, usually into or near streams and rivers, with little or no attention paid to its effect.

Many of the early gold prospectors mined by panning, which involves swirling gravel and sands scooped from stream beds in a shallow metal pan to concentrate the denser gold particles. The less-dense waste material rises in the pan and can be washed away. Another placer-mining technique involved pouring the stream gravel into a sluice — a long trough that contains a series of upraised riffles along the bottom. The denser gold particles are trapped in the bottom by the riffles, while the less-dense sediments are washed away. Panning and sluicing are simple forms of placer mining — mining for minerals in sediments that form when underground lode deposits, ore deposits in solid rock, erode and the eroded particles are deposited by water in beds. These centuries-old low-capital mining methods are still practiced today by small-scale independent miners in the United States and around the world, but they depend on a great deal of low-cost labor per unit of product.

As the amount of California gold which could be recovered easily by stream panning dwindled, a new, more capital-intensive method of placer mining began to be practiced. Commonly called hydraulicking, it was first used in Nevada County, California, in the 1850s. Hydraulicking entailed the spraying of river gravel banks with pressurized water and capturing the runoff in long sluices to recover the denser gold particles.

The high-pressure nozzles used by hydraulic operations spewed out water at the rate of 185,000 cubic feet per hour, washing out huge chunks of hillside and leaving the land permanently scarred.[15] During this process, most of the dislodged material was flushed through the sluices and washed away in streams. Successful hydraulic mining required vast and readily accessible quantities of water. To obtain this, mining companies dammed up rivers to ensure supply and built over 5,000 miles of

PHOTO: LIBRARY OF CONGRESS

Hydraulic mines, like this one in Idaho, sprayed water at extremely high pressures against gravel banks to loosen mineral-rich ores. Wastewater washed away into nearby streams, creating long-term pollution problems.

water delivery systems such as canals and pipelines to transport the water from reservoirs to their mining sites.[16]

The large amount of sediment which was mobilized by hydraulicking choked natural streambeds with mud and sand. As the original watercourses became filled, flooding became more frequent and more severe. On a number of occasions, for example, waters from the swollen Yuba and Feather Rivers flooded the important agricultural town of Marysville, California, which lay downstream of hydraulic mining areas. The town had to erect high levees to protect itself.[17] The floodwaters also deposited a yellow mixture of mine waste, in the form of sand and silt, from hydraulicking operations. Known as slickens, this waste smothered agricultural land and its crops.[18] Moreover, the silt-laden waters were unsuitable for livestock, irrigation, and other uses. The fishing industry also suffered as well from this form of pollution.[19]

Outraged agricultural interests soon mobilized to stop hydraulicking. They initiated a series of lawsuits against mining companies. One of the most significant cases, *Woodruff v. North Bloomfield Gravel Mining Company et al.*[20], was filed in 1882 after a series of disastrous hydraulicking-induced floods on the Yuba River in Marysville. In 1884, Judge Alonzo Sawyer of the U.S. Ninth Circuit Court ruled in favor of farming interests, enjoining (or stopping) the North Bloomfield Gravel Mining Co., a hydraulicking operation, from discharging debris into the Yuba River or its tributaries. What can be considered one of the Nation's first environmental regulations for mining — the "Sawyer" decision — marked the beginning of the end of the destructive hydraulic mining period in California. Because of the Sawyer decision, other courts began issuing decisions that stopped other damaging hydraulicking operations across the country.[21] During the 30-year hydraulic period, gold mining in California surged, producing an average of $10 million worth of gold per year.[22]

In 1893, the U.S. Congress passed the "Caminetti Act," which provided for restricted hydraulic mining in the California region under the tight control of the California Debris Commission. The Caminetti Act authorized the Commission to issue permits for hydraulic mining in territory drained by California's Sacramento and San Joaquin Rivers. However, the law required hydraulicking operations to impound all their debris so that the wastes would not settle on valley floors. The costs of complying with this requirement ensured that only a handful of companies engaged in the practice — and only on a limited basis.[23] Hydraulicking did continue in some states, such as Montana and Alaska, but under the tight controls mandated by the earlier Sawyer decision.

With hydraulicking's demise, mining for lode deposits — hardrock ore bodies in their native rock — typically by underground methods, dominated metal production in the U.S. Ores from lode deposits were increasingly handled in industrial-sized mills equipped to grind and concentrate thousands of tons of ore per day. To lower production costs, smelters, ovens used to melt ores in order to separate the desired metal from waste material, were often built close to mining areas.

During the late 19th century, major mining and processing centers developed throughout the Western United States: Virginia City, Nevada; Leadville, Colorado; Bingham Canyon, Utah; Butte, Montana; Globe and Miami, Arizona; Bunker Hill, Idaho. Mining booms during this time also took place in the Keweenaw Peninsula in Michigan, and in southeastern Tennessee. As mining output grew, waste production and water pollution impacts grew as well. Mining of the Comstock lode near Virginia City, Nevada, flourished from the 1860s through the 1880s. Though the Comstock lode, discovered in 1858,[24] was first mined for gold, it yielded one of the world's great fortunes in silver and turned Virginia City, Nevada into a boom town. At one point the Comstock district supported over 60 major active mines. Mining production peaked in the late 1870s. In 1877, two mines, the Consolidated Virginia and the California, produced over $30 million in metals.[25]

Gold was discovered in the Leadville, Colorado, area in 1859. In a succession of booms and busts, silver, lead, and zinc were produced from area mines. By 1927, it was estimated that over 75 miles of tunnels and shafts had been bored to access the area's ores and drain underground mines.[26] The Leadville district became one of the most prolific mining areas in the country, eventually producing precious metals worth more than $2 billion.[27]

Discoveries of copper also resulted in mining booms. In 1882, rich veins of copper were found near Butte, Montana. Aptly named the "richest hill on earth," the Butte district yielded over 13 billion pounds of copper in over one hundred years of mining.[28] Copper mining on Michigan's Keweenaw

Peninsula, which began in the late 1860s, has produced over 10 billion pounds of copper.[29]

Though mining activity has declined greatly in most of these areas, the legacy of these "boom and bust" mining centers is still evident. The mills, smelters, and mines built in the last century have left behind poisonous wastes which continue to damage aquatic life and place human health at risk.

At the Comstock lode, "stamp" mills crushed silver and gold ore into a fine powder. Mercury was added to the ground ore to extract the precious metals, and the leftover contaminated tailings were discarded in nearby streams.[30] In Butte, mines dumped tailings directly into the Clark Fork River from the beginning of mining to about 1910. Later, tailings were impounded in hastily-built ponds which now flush fresh heavy-metal laden muds into the river after every heavy rain.[31] On Michigan's Keweenaw Peninsula, copper mills discharged an estimated 200 million tons of copper-contaminated waste directly into Torch Lake, reducing its volume by 20 percent and leaving a toxic threat to fish and anything that eats them.[32] And in the Copper Basin area of southeastern Tennessee, 32,000 acres of land were stripped of trees, mostly to provide fuel for copper smelting and open-air roasting of copper ore. Dense sulfurous smoke from the roasting blanketed the hills. The result was a lifeless landscape and severe, long-term erosion that has continued to clog Tennessee's Ocoee River with metal-contaminated sediments and has devastated most of its fish life.[33] Over fifty of these mining operations, started in the 19th century, have become today's hazardous waste Superfund sites.[34]

Mining Methods in the 20th Century

Today, surface mining accounts for approximately 97 percent of the ore tonnage extracted by hardrock mining in the United States.[35] Underground mining, once dominant, has dwindled considerably since the beginning of the century, while the volume of surface-mining extraction has ballooned. The ascendancy of surface mining and the development of new technologies for moving vast amounts of earth and for extracting metals from lower grade ore have created enormous quantities of new and dangerous mining waste.

Surface Mining

Surface mining can be visualized in two general forms, though they overlap. The two are strip mining and open-pit mining. Strip mining consists of stripping away layers of soil and waste rock over a mineral deposit. It is the term often used to describe mining for minerals like coal or phosphate that are found in relatively horizontal geologic deposits formed in layered seams. Strip mining leaves cuts in the earth where waste material and minerals have been removed. After the minerals are extracted, the cuts can be backfilled with the previously-removed layers of waste rock and soil.

PHOTO: PHILIP M. HOCKER/ MINERAL POLICY CENTER

Surface mines, like the massive open-pit shown here (left center) at Bingham Canyon in Utah, dominate hardrock mining in the United States today.

Open-pit mining involves excavating the earth's surface in a concentrated location to access the underlying mineral ore body. Open-pit mining is the common description for surface mines that access metal ores, such as copper, gold, and silver. Metal ore deposits are often found below the surface in bodies that have varied, irregular shapes and concentrations, and are surrounded by large amounts of waste rock material. To reach these deposits, the pit must be dug in a progressive series of stages. The walls of open pits are usually terraced, or "benched" — flat steps, perhaps 15 feet wide, are left between near-vertical walls, which may be 40 to 60 feet high. This benching is done for safety, in order to avoid massive landslides due to collapse of the rock on the walls, and to provide level working surfaces for mining machinery. Because the walls step back, each increase in the depth of the pit normally requires an increase in its width all the way to the top. Eventually, the increase in cost of moving the walls back becomes so high that mining to further depth is unprofitable, and the mine is shut down.

Open-pit mines can be extremely large — sometimes more than a mile across and several thousand feet deep. For example, at the Kennecott Corporation's Bingham Canyon Mine near Salt Lake City, Utah, the world's largest open-pit copper mine, the pit measures one-half mile deep and almost 2.5 miles across.

Open-pit mines create enormous quantities of waste rock, which is usually stored on the earth's surface in vast piles which may reach hundreds of

Rail cars at underground mines transport ore. Underground mining is no longer a dominant hardrock mining method in the United States, although some lead, zinc, gold, silver, and copper are still extracted this way.

PHOTO: PHILIP M. HOCKER/ MINERAL POLICY CENTER

feet high. Open-pit operators almost never put this waste back into the pit when the mine closes. Mining companies claim that backfilling pits is a prohibitively expensive reclamation practice. Another reason mine operators are reluctant to backfill is that it would economically prohibit a company from reopening the pit sometime in the future. If mineral prices rose or new more-efficient mining technologies became available, a played-out pit with marginal ore deposits left below the pit's floor might become profitable to reopen. Filling the pit with waste rock would substantially raise the cost of reopening the pit.

Underground Mining

Underground mining shares many of the techniques of surface mining. Instead of carving a huge pit in the earth, however, underground mine operators dig shafts (vertical tunnels for access and ventilation) and adits (horizontal tunnels for access and drainage) to reach the ore. The ore extracted below ground is carried to the surface through the shafts and adits by truck, rail car, conveyor, or skip (an elevator used to transport ore).

Underground mining is much more expensive per ton of material mined than surface mining. More highly-skilled workers are needed, along with specialized equipment. The scale of equipment and productivity rates are lower. For these reasons, underground methods are only used where (i) surface mining is impractical — usually because the ore body is deep below the surface — and (ii) the ore body has a sufficient value per ton to pay for the high costs. Underground mining generates much less waste than surface methods. Because it costs as much to mine a ton of waste rock as to mine a ton of ore, and because the per-ton costs are quite high, there is a powerful incentive to minimize the amount of waste extracted.

Lead and zinc are the only major metals extracted today primarily by underground mining.[36] However, small but significant amounts of gold, silver, and copper are also mined by underground methods. In the field of non-metal mining, approximately 90 percent of salt is obtained by underground mining, either in rock form from underground room-and-pillar mines or by injecting fresh water underground and collecting the salt-bearing liquid or brine through extraction wells.[37]

As in surface mining, waste rock displaced by underground mining is usually stored on the land surface. However, some underground mines are "backfilled" — the waste rock or tailings are replaced to fill the underground cavities the mining has left. Backfilling helps prevent subsidence of the land over the mine, and reduces the amount of waste that must be stored aboveground. Backfilling has its own risks, however: if the material put back into the workings is acid-forming or leaches contaminants, it can contaminate groundwater that percolates through the mine.

Mineral Processing

Regardless of what mining method was used to extract the ore, it must be processed to separate the target mineral from the rock ore and to remove non-valuable metals, soil, and other impurities.

Ore is usually crushed before it undergoes beneficiation to separate and recover the target mineral. The beneficiation process employed will vary depending on the desired product. Some mineral products, such as clays, marble, limestone, and sand, require relatively little beneficiation because they are useful in roughly the same form in which they are extracted. In contrast, the beneficiation of metals such as copper, gold, zinc, and lead involves much more processing. These metals appear in small particles or chemical compounds throughout the ore; therefore several processing steps must be taken to convert the raw ore to a more usable, concentrated form.

One of the most common methods of beneficiation is known as flotation. Flotation methods are used to concentrate metals like copper, lead, and molybdenum. The first step in this process consists of crushing and grinding the ore to a powder. Next, the powdered ore is mixed with water to form a slurry and is pumped into a series of open troughs called flotation cells. Chemical reagents known as collectors and depressants are usually added to the slurry to separate the target mineral from the non-valuable material, causing one to either float to the top or sink to the bottom of the cell while the other material goes the other way. In a typical flotation process, the addition of collectors causes the mineral particles which contain the target metal to attach themselves to air bubbles injected into the flotation cell. The particles are carried to the surface by the bubbles and then skimmed off the top as a "concentrate," while the non-valuable residue, known as gangue, is collected below.

Processing reagents used in the flotation process include cyanide, fuel oil, kerosene, amines, sodium sulfide, sodium carbonate, ammonia, and lime. These chemical reagents can be highly toxic. Cyanide, for example, is highly toxic to humans, and can kill fish and wildlife in very low doses.[38] Amine chemicals and sulfuric acid, also used in flotation, can be extremely irritating to the respiratory tract and mucous membranes.[39]

In addition to flotation, other means of beneficiation include gravity concentration and magnetic separation. Gravity concentration, as the name implies, separates minerals based on differences in their "gravity," or density. Gravity concentration is the traditional method of separating gold from gravel material — just as early prospectors used gravity in sluices to recover gold particles. Magnetic separation is commonly used to separate iron ores, which often exhibit magnetism, from less magnetic material. Some newer beneficiation processes, such as chemical leaching, are discussed in the next section in greater detail.

Regardless of the separation method, however, waste is the chief byproduct of the beneficiation process. The finely ground milling wastes left over from beneficiation, known as tailings, consist of pulverized ore and chemical residue. Tailings usually wind up in a slurry form mixed with mill processing water. Frequently, tailings are placed in ponds or impoundments at a site. The liquid portion of the tailings waste then slowly evaporates, or it seeps into the groundwater. Eventually, the semi-liquid tailings waste dries into piles or beds of dry, fine rock powder. Tailings are the second greatest form of mine waste, constituting close to one third of all mine waste generated.[40] In gold and copper mining, specifically, tailings constitute approximately 45 percent of all material handled.[41]

Smelting is a subsequent mineral processing step in which intense heat in a smelter furnace is applied to the beneficiated ore concentrate. The purpose of smelting is to extract the metal from the metal concentrate. Smelting melts the metal, removes impurities such as sulfur and other metals, and allows the purified metal to be poured into ingots or slabs. Until very recently, smelters have been notorious polluters. If operated without air-pollution controls, smelters release vast quantities of sulfur dioxide and other chemicals into the air. Sulfur dioxide is a major source of acid rain. Smelter furnaces also produce a glass-like solid waste known as slag, which consists of the cooled, solidified impurities which have been removed from the product. Slag contains a variety of metals such as iron, copper, lead, cadmium, arsenic, and barium. Slag is often stored on the earth's surface in piles.[42] Though it appears glass-like, the metals it contains can leach out of slag. When not properly protected from the elements, slag provides yet another source of mining-derived water pollution.

Many metals are further refined to a higher degree of purity with electricity. Known as electrorefining, this process is commonly used to refine copper, lead, and zinc. This process also leaves concentrated chemical wastes, known as slimes. Slimes, which can contain hazardous heavy metals, often are resmelted or refloated to recover gold and silver. The remaining portion of the slime, after precious metals are removed, is usually disposed of on land.

New Chemical Mining Methods

Minimizing the handling of ore is one of the best ways to lower production costs at a mine. An important method for doing this is to use chemicals to treat the ores without first crushing and grinding them — or maybe without digging them out of the ground at all. Generally, this group of methods is called "chemical" or "solution" mining, because it relies on the ability of specific chemicals to mobilize target metals in liquid solutions, while the non-marketable residue is left behind as a solid.

Solution mining has been around for a long time, but its recent expansion and sophistication has revolutionized the hardrock mining industry. It is much cheaper to process a ton of ore with solution mining than with conventional methods of grinding and flotation, which require large investments in expensive machinery and large expenditures of energy. The use of solution mining has given the mining industry a way to profitably mine low-grade ore deposits which would otherwise be uneconomical. Consequently, solution mining has made it profitable to mine large amounts of ore which would have been left unmined in the past, and has led to the creation of new landscapes of heaps of spent ore left after solution mining is completed. Today, solution mining accounts for a significant portion of the nation's gold, silver, and copper production. The primary methods of solution mining are *heap, vat, and dump leaching.*

Gold, Cyanide, and Heap Leaching

The chemical cyanide (in ionic form, CN^-) is effective as a leaching agent because it forms a soluble bond with gold (Au) and silver (Ag). In most leaching operations today, the cyanide ion is supplied by dissolving sodium cyanide (NaCN) in water with an elevated pH; the solution is then used as a leaching agent. When a solution of sodium cyanide is exposed to gold and silver ores, a chemical reaction occurs, forming soluble complexes of gold ($NaAu(CN)_2$) and silver($NaAg(CN)_2$).[43] The cyanide-gold and cyanide-silver complexes are dissolved into the leaching solution and are then separated from it by further processing.

The use of cyanide to separate gold from ore was first developed in Scotland in the late 19th century. This method was an improvement over the inefficient mercury amalgamation process that was used at the time, which usually recovered no more than 60 percent of an ore's gold value. In contrast,

PHOTO: PHILIP M. HOCKER/ MINERAL POLICY CENTER

A toxic cyanide solution is sprinkled atop heaps to extract gold and silver from the ore at this Nevada mine. The metal-rich solution is collected in pools (lower right).

leaching finely-ground gold ore with cyanide could recover more than 97 percent of the ore's gold value. Adopted quickly by industry, the new leaching process first saw widespread use in the Witwatersrand gold fields of South Africa in the late 1800s.

Cyanide heap leaching is an outgrowth of these early methods of processing gold ore with cyanide. Heap leaching with cyanide was proposed by the U.S. Bureau of Mines in 1969 as a means of extracting gold from ores which had been considered too low in value to process economically. The gold industry adopted the technique in the 1970s, soon making heap leaching the dominant method of treating gold ores. Silver is a secondary product of heap-leaching many gold ores.

In cyanide heap leaching, ore is extracted and piled in heaps often hundreds of feet high. Sometimes the ore is first crushed to a smaller size to increase the efficiency of the leaching process. A dilute sodium cyanide solution — usually 250 to 500 parts per million of sodium cyanide — is then sprayed on the ore heap through an irrigation system and allowed to trickle through it. As the cyanide solution percolates down through the heap, gold particles in the ore bond with the solution. The resulting fluid, known as "pregnant solution" because it contains the cyanide-gold (or cyanide-silver) complexes, drains down a synthetic liner to the bottom of the pad. The solution then flows over the liner to the perimeter of the heap where it is collected and pumped by pipe or drain to a lined retaining pond. Collection ponds and heap pads are usually lined with so-called impermeable material (usually plastic over clay, sometimes asphalt) designed to avoid the loss of the gold-bearing liquid and to prevent cyanide solution from escaping into ground or surface water.

Heap leach piles are usually constructed in separate lifts, or layers of ore. After most of the gold has been leached out of one lift, a new lift of fresh ore is placed on top of the old lift. The heap is again leached with solution. Each leaching cycle may take up to three months. After a sequence of leaching cycles, piles can tower as high as 300 feet. Some heaps sprawl over hundreds of acres.

Once the gold-bearing liquid is collected in the pregnant solution pond, it is pumped to a processing facility for recovery of the metals. At most U.S. gold mines, gold is then recovered by first running the pregnant solution over activated carbon materials (usually roasted coconut shells), which retain the gold-cyanide complex. The carbon is then washed with a highly alkaline cyanide solution to release the gold. The gold is recovered by adding zinc to the wash solution, initiating a chemical reaction that causes the gold to precipitate out (to settle out of a solution by gravity). The resulting impure gold is usually further purified by electrical refining and smelting.

The cyanide solution that is left over from processing is stored in outdoor ponds and then re-used on the heap in a subsequent leaching cycle. Fresh cyanide is added to this solution before another round of leaching. Heap leaching requires large amounts of cheap sodium cyanide. Hundreds of millions of pounds are consumed each year by the mining industry.

A similar heap leach process is used in some copper mines. Like gold, copper ore is extracted and stacked in heaps. The copper ore heap is sprayed with a sulfuric acid solution, rather than cyanide, to separate and mobilize the copper from the ore. The pregnant solution is collected and processed to remove copper from the solution (more detail below in discussion of heap and dump leaching).

While heap leaching involves intricate steps to remove the gold, silver, or copper from the ore and chemical solution, it substantially reduces the amount of ore handling required by traditional mining methods. The more times and the greater the distances over which rock ore is moved and processed, the higher the production costs. With heap leaching, once the ore is crushed and stacked in heaps, no further ore handling occurs.

Gold, Cyanide, and Vat Leaching

Cyanide was first used to extract gold from ore in large tanks, or vats, which led to the term "vat leaching." With vat leaching, ore is finely ground in large circular mills and piped into large tanks or vats. A cyanide solution is added to the powdered ore. The same chemical reaction occurs between gold and cyanide as in heap leaching, and the solution bearing the gold is separated from the ore residue. The residual powdered rock is now tailings. The tailings are piped out of the mill to be disposed of in large ponds.

Because vat leaching is applied to a pulverized ore, more of the cyanide solution makes contact and can bond with the gold (or silver) in the ore than would be the case in heap leaching. Consequently, this method recovers a higher percentage of the valuable metal which is in the ore, typically over 90 percent. The miners must decide whether there is enough gold concentration in the ore to pay for the higher cost of vat leaching. This decision is made on a truckload-by-truckload basis at today's large open-pit gold mines. Although more tons of gold ore are heap-leached than vat-leached in the U.S. today, a greater quantity of gold is actually produced by vat leaching, because the method is used on higher-grade ores and has a higher gold recovery rate.

Whether by heap or vat method, cyanide leaching's main advantage is that it is cheap, allowing for economical recovery of even minute concentrations of gold. Heap leaching, in particular, has revolutionized gold mining in the 1980s and 1990s, and the industry has experienced a new boom centered in northern Nevada's Carlin Trend—an extensive area of low-grade gold deposits. In 1989, cyanide heap leaching produced 3.7 million ounces of gold from 129.8 million tons of ore. That same year, cyanide vat leaching produced 4.3 million ounces of gold from 40.6 million tons of ore.[44] Cyanide leaching is now used to process over 90 percent of gold ores by weight.[45]

Copper Heap and Dump Leaching

Copper mining uses a chemical leach process similar to gold leaching. For copper, however, dilute sulfuric acid is used as a leaching agent instead of cyanide.

In the past, the grade of copper ore was too low to pay the cost of milling and it was disposed of in vast dumps, piled directly on the ground with no protective liner underneath. Leaching these dumps with sulfuric acid is known as dump leaching. However, today copper ore is typically heap leached — crushed and placed on a pad or impermeable lining before being soaked with the leaching liquid. Like gold leaching heaps, copper leach heaps are constructed in successive lifts or layers. They often grow to enormous dimensions, spreading over hundreds of acres and reaching heights up to 200 feet.[46] Copper-bearing solution is collected at the bottom of the heap and is pumped to pregnant solution ponds for further processing.

Copper leaching dates back to 18th-century Spain when miners in the Rio Tinto area discovered they could recover copper dissolved in the runoff from ore dumps. Modern copper leaching is combined with a technique called solvent extraction/electrowinning ("SX/EW") to produce a highly refined copper product ready for use in manufacturing.

In SX/EW, after the copper-bearing pregnant sulfuric acid solution is collected from the leach piles it is routed to a solvent extraction tank. In this tank, copper is extracted from the pregnant solution using an organic copper extractant (usually composed of aldoxime or ketoxime compounds) which is usually dissolved in a kerosene base.[47] The organic agent selectively bonds with the copper, leaving behind other metals and non-metal impurities in the

PHOTO: PHILIP M. HOCKER/ MINERAL POLICY CENTER

Vat-leaching, using cyanide solution, is a cost-effective method for recovering gold from ores. Ore is ground finely, loaded into the vats, then mixed with cyanide solution to extract the gold. Tailings are disposed of in large ponds.

original pregnant solution. Sulfuric acid is added to the extracted copper, forming an electrolytic (electric-conducting) solution. The electrolytic solution is then conducted to electrowinning cells or tanks. An electric current is sent through the electrolytic solution, causing the dissolved copper to deposit onto metallic plates, or cathodes. The deposited copper produced by the SX/EW process is so pure (99.9 percent copper) that it needs no further smelting or refining before it is ready for manufacture.

The SX/EW process now accounts for approximately 27 percent of U.S. copper production.[48] Tighter controls on air pollution from smelters have made SX/EW methods more attractive to the industry. However, higher grade copper ores are still processed using conventional beneficiation and smelting methods.

Chemical Mining "In-Situ"

The Latin phrase "in-situ" means "in position." In-situ mining is another form of solution mining, but it is fundamentally different from conventional surface and underground mining techniques. With in-situ mining, the ore is not removed from the ground and processed. Instead, the ore is left intact in the ground, and chemical extractants are passed through the ore in place and collected to recover the minerals. In-situ mining is being used to extract copper, uranium, and some non-metals such as potash and soda ash.

In in-situ mining, a chemical solution called a lixiviant is injected underground into an undisturbed ore deposit through a series of pressurized injection wells. The solution percolates through the rock, mobilizing or bonding with the target mineral. A production/recovery well some distance away then draws up the liquid to the surface for subsequent recovery of the metal. Not every type of rock is suitable for this mining method. The ore deposit must be sufficiently permeable to allow the solution to circulate through the rock and come in contact with the target metal.

In-situ mining has also been used as a supplementary mining method, in conjunction with conventional surface or underground mining, to recover mineral residues remaining at a mine site that are no longer economical to recover through conventional mining methods. Since the early 1940s, in-situ mining has been used in Arizona to extract copper left in old underground mine workings. In a spent underground mine, operators often blast the pillars of ore left holding up the old underground mine chamber ceiling. Lixiviant is then injected by wells into the chamber rubble where it percolates through the remaining ore body, and then is pumped to the surface through production/recovery wells to recover the remaining mineral residue in the underground rubble. This process is also used in spent open pit mines, where injection wells pump the lixiviant into the pit's benches (terraced levels). The solution percolates down through the rock and is collected at the bottom of the pit through a sump pump and transported for final processing.

In-situ mining makes moving rock and piling it in vast heaps unnecessary. Because this method virtually eliminates earthmoving, it is a much cheaper method of extracting minerals than conventional mining. If done properly with complete control of injected materials, it offers environmental advantages: in-situ mining generates no tailings, greatly reducing solid waste, and leaching solutions can be recycled continuously.[49]

However, in-situ mining is also fraught with environmental risk — mainly because predicting the movement of a lixiviant underground is very difficult. Of greatest concern is the possibility that a lixiviant may deviate from its projected path and contaminate ground water. The following examines in more detail the in-situ mining method for copper and uranium.

In-Situ Copper Mining

To date, in-situ copper mining in the U.S. has not been practiced commercially except as a supplementary, "mop up" phase of a conventional mine. Recently, however, some copper companies have shown interest in in-situ copper mining as an inexpensive method to extract copper from very low-grade, previously undisturbed ore deposits.

BHP Copper (a division of BHP, an Australian-based mining conglomerate) is now developing the first commercial-scale in-situ copper mine near Florence, Arizona. In 1996, the Florence Project was still in the environmental permitting stage. The project proposes injecting a weak sulfuric acid solution into bedrock 500 to 1,200 feet below the surface to extract low-grade copper. As the solution migrates through the permeable ore body it will mobilize the copper. The copper-laden solution will be drawn up to the surface through a series of production/recovery wells and then processed to produce copper using the SX/EW method.[50] BHP will also sink monitoring wells outside of the ore body to detect if the sulfuric acid solution migrates beyond the recovery wells.

BHP's project follows closely on the heels of an experimental in-situ copper mining project still underway near Santa Cruz, Arizona. The Santa Cruz project, a cooperative project between ASARCO Inc., Freeport-McMoRan Copper & Gold Co., and the U.S. Bureau of Reclamation, has already begun extracting copper from a low-grade ore body using a sulfuric acid solution.

Prior to being abolished by Congress in 1995, the U.S. Bureau of Mines had been a partner in the Santa Cruz project. The Bureau promoted in-situ leaching as a promising means of extracting low-grade copper cheaply and with minimal environmental impacts.[51] The Bureau of Mines' role in this Federal-private partnership has since been assumed by the Bureau of Reclamation. But this enthusiasm may be premature. Operators must maintain rigorous control over lixiviants because the movement of fluids underground is often difficult to predict. There is a risk, too, that acid lixiviants could dissolve other toxic metals, escape through rock fractures, and contaminate drinking water supplies. The former Bureau of Mines, an ardent supporter of the mining industry, had also suggested that cyanide be used for in-situ gold leaching.[52] To date, in-situ gold mining with cyanide has not been attempted. Given cyanide's acute toxicity, an operator miscalculation with this method could produce even more serious harm to groundwater resources than would sulfuric acid.

Even assuming that in-situ copper mining offers environmental advantages, it is not likely to be practiced broadly. Rather, in-situ mining is likely to be favored only where very low-grade copper ores are found in permeable rock. Conventional extraction, flotation, milling, and smelting are still the favored methods for mining higher grade copper ores because of these methods' greater efficiency in recovering copper. Traditional flotation and smelting methods can recover more than 90 percent of copper contained in ore, while solution mining, such as in-situ mining, usually does not recover more than 75 percent of copper in ore.[53]

In-Situ Uranium Mining

While in-situ copper mining of undisturbed ores is still in its infancy, in-situ mining of uranium has a more developed history. Pioneered by the Atomic Energy Administration, in-situ mining has been practiced commercially since the mid-1970s to extract radioactive uranium from porous sandstone aquifers in Wyoming and Texas. In-situ mining now accounts for a majority — almost 70 percent — of United States uranium production. Another 27 percent of total uranium production comes from the White Mesa mill in southeastern Utah, which processes previously stockpiled ore left by now closed surface uranium mines.[54] The small remaining portion of uranium production occurs as a byproduct of the processing of phosphate rock, mined mostly in the Eastern United States. As of 1996, there are no active surface uranium mines in the United States.[55]

Current in-situ uranium sites use alkaline solutions consisting of sodium carbonate-bicarbonate, carbon dioxide, water, and oxygen as leaching agents to mobilize uranium. The leaching solution is injected into the ore body through a pattern of injection wells surrounding a central production/recovery well. The solution mobilizes the uranium as it migrates through the ore body. The uranium-bearing solution is drawn through the production/recovery well, and then piped to a recovery plant where the uranium is processed and removed.[56]

In-situ uranium mining can present environmental risks similar to those posed by in-situ mining of copper. The leaching agents can mobilize uranium, other radioactive constituents, and heavy metals that can escape to areas outside the mining operation. These contaminants may migrate into useable groundwater zones and neighboring aquifers. Although regulatory agencies often require operators to conduct groundwater monitoring, and may impose a requirement to restore aquifer water quality to background levels after mining, there is still uncertainty as to how effectively operators can control leaching solutions. Since most in-situ projects have been located in remote areas, so far there have been no reports of drinking water supplies contaminated by in-situ uranium operations.

Because of a poor market for uranium over the last two decades, only four in-situ uranium mines remained in operation in the U.S. as of 1994.[57] At that time, thirteen other in-situ operations were on inactive standby or undergoing decommissioning. However, a rise in uranium prices may breathe new life into this mining method. One proposed in-situ project is located near the town of Crown Point, New Mexico. Hydro Resources, Inc. is proposing to operate several in-situ uranium mines at this location on the edge of the Navajo Nation Reservation. Crown Point's residents are particularly concerned about the possibility that these mines may contaminate the area's high quality groundwater. Wells used for the community's drinking water and for livestock are located in the same aquifer, only one-half mile away from the proposed project.

Mining: A Wasteful Business

Regardless of the mining process used, mining is a startlingly wasteful business. Mining is more waste-intensive than any other natural resource extraction industry. Forestry converts more than 50 percent of its harvested wood into marketed wood or paper products. The processing of crude oil into fuel and petrochemicals produces a relatively small amount of waste.

Mining, on the other hand, generates much more waste than product. In iron production, only about one third of all the ore extracted and processed (by weight) becomes a final refined product. This leaves two thirds of the material as waste. Surprisingly, however, iron mining is one of the least wasteful

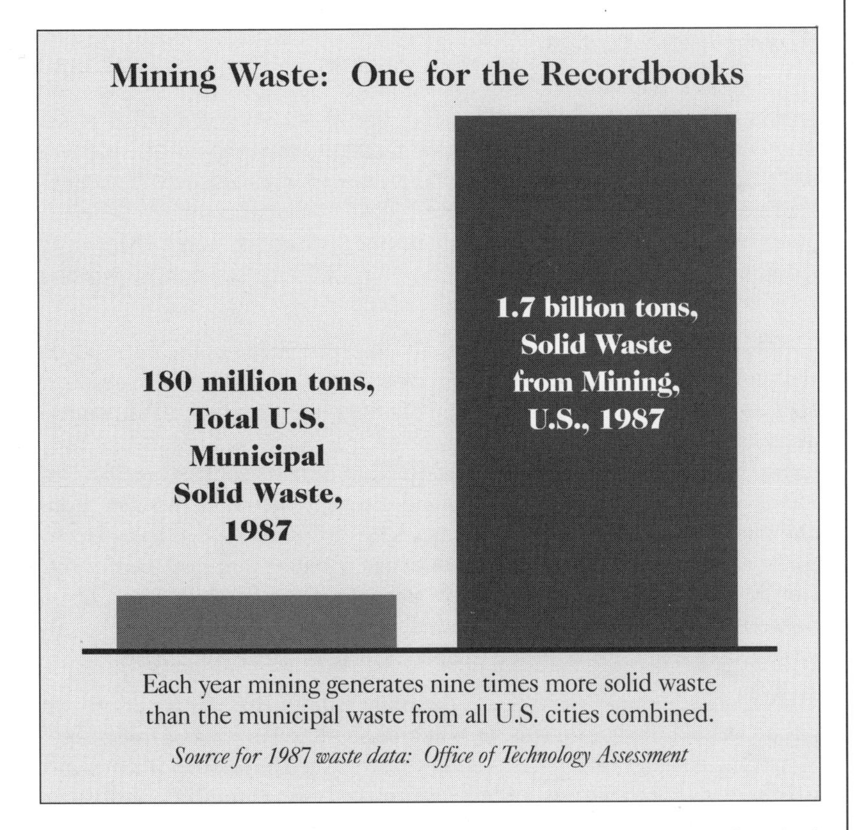

Each year mining generates nine times more solid waste than the municipal waste from all U.S. cities combined.

Source for 1987 waste data: Office of Technology Assessment

1.7

Billion tons, Solid Waste from Mining

180

Million tons, Total U.S. Municipal Solid Waste

hardrock mining processes. By contrast, of all the materials used and disturbed in the lead mining process, only 5 percent becomes a final product — leaving 95 percent in mine waste material. Refined copper represents only 0.6 percent total of raw materials used and disturbed in copper mining. Silver yields only 0.03 percent.[58] Of the major metal mining industries, however, gold mining is the most waste intensive. Refined gold represents only about 0.00015 percent of all the raw materials used in the gold-mining process.[59] In fact, *it takes about 2.8 tons of gold ore to produce just one man's wedding band.*[60] The rest, of course, is waste.

Although copper mining generates less waste per ton of ore mined than mining for other metals like gold, it generates the greatest total amount of waste by weight of all forms of metal mining, because of the huge volumes of copper mined.[61] In 1992, for example, U.S. copper production generated approximately 807 million tons of waste, while gold production generated approximately 596 million tons.[62] Copper waste constitutes approximately 45 percent (by weight) of total metal mining waste generated in the U.S. Gold production contributes a 17 percent share of total U.S. metal mining waste, while iron ore contributes about 10 percent.[63] Mining for other metals such as lead and molybdenum generates smaller, but still significant, amounts of waste.

Among non-metal minerals, phosphate mining generates the greatest total amount of waste (by weight). Phosphate is one of the principal raw materials used in the manufacture of fertilizers (see Phosphate Mining Waste in this chapter).

Mining Waste Sources

Regardless of the mineral being mined, enormous quantities of waste are generated by the extraction and beneficiation of ore. As noted earlier, much of this waste consists of waste rock and tailings. Waste rock is the rock and soil that does not contain the target mineral but must be removed in order to gain access to the mineral deposit (ore). Waste rock may contain acid-generating materials and metals that can be transported by water. Moreover, stormwater washing over these piles may carry off large amounts of rock and soil that can clog rivers and lakes with sediment.

Tailings comprise ground-up ore and the processing chemical residues that remain after the ground ore is processed to remove the salable product, such as gold, copper, or molybdenum. Tailings often contain sulfide minerals, such as iron pyrite, which can generate acid mine drainage when they come into contact with water and air. Tailings can also contain metallic elements of various toxicity — including arsenic, cadmium, mercury, lead, chromium, and zinc — as well as processing chemicals such as cyanide and sulfuric acid. Because of their fine consistency, tailings are easily carried off-site by wind or water. If improperly stored, tailings can spill, leach, or be washed or blown off-site into surface and groundwater resources, causing serious environmental damage (see Chapter 3, Erosion and Sedimentation).

Nationwide, more than half of the mine tailings that are generated are placed behind dams in ponds, or waste impoundments, at the mine site.[64] Sometimes tailings material is used in constructing the impoundment dams themselves. In other cases, tailings are disposed of in natural depressions, or in the open pits left by the mine's operations. A very small portion of tailings is used offsite in building materials and road surfaces. Overseas, some mining operations continue to dump their tailings directly into rivers to be washed downstream.

The key to preventing environmental problems from mine waste is two-fold. First, mine operators must minimize the amount of waste that is generated. Second, operators must take adequate measures to ensure that mine waste is properly and securely stored and isolated from natural elements, such as air and water, that can mobilize and transport contaminants in the waste. Wastes that can react to form acid must be managed to prevent the acid-forming reactions from occurring. Although many mining pollution control practices have improved over the years, mine sites today continue to experience break-downs that cause tailings impoundments and waste rock piles to spill, leak, or discharge pollutants into groundwater and surface waters. The following sections examine various factors relating to mine waste storage and disposal.

Tailings Impoundments

Most of the early metal mines in the United States dumped tailings directly into streams. As stream water washed over the tailings, it would carry a portion of the mine waste downstream, allowing mine operators to dump more tailings in the stream. Damage from this practice was extensive. The Clark Fork River complex near Butte, Montana, now the scene of this country's largest area of related Superfund sites, is a prime example of environmental damage caused by tailings disposal into rivers and streams.

From the 1860s to the early 1900s, gold, copper, and silver mines and processing facilities in the Butte area discharged mine waste onto surrounding land and into streams. In all, mining operations deposited approximately two million cubic meters of contaminated sediments.[65] The Clark Fork's floodplain became a wasteland, encrusted with this contaminated material. Even today, rainstorms, snowmelt, and floods continue to wash mine tailings into the Clark Fork and its tributaries. The contaminated portion of the Clark Fork extends 120 miles downstream from Butte. Cleanup costs are expected to reach at least $320 million for the three contiguous Superfund sites related to Butte's mine waste[66] (see Chapter 7 for more detail).

As major mining regions prospered and flourished in the early 1900s, so did agricultural development. Uncontrolled mine waste disposal soon led to conflicts with agricultural interests over water sources, water use, and mining contamination. Tailings that washed down the rivers clogged up irrigation channels and polluted irrigation water. These incidents resulted in litigation between agricultural and mining interests. By about 1930, these conflicts generally led to the end of uncontrolled tailings dumping. Thus began the modern era of tailings impoundments, though some mines continued to flush tailings into public waterways until Clean Water Act lawsuits forced them to stop in the 1970s.[67]

Today, mining companies operating in the U.S. are no longer permitted to dump tailings into rivers. Instead, mine operators often store tailings in a pond or impoundment behind an earth-fill dam. Early dam construction was crude and often highly unstable, but with the advent of large industrial earth moving equipment in the 1940s, and improved methods of engineering design, tailings dam construction has improved dramatically. Some of today's impoundments are lined with sealed sheets of theoretically impermeable synthetic materials to prevent tailings from leaking into surface water and groundwater. Tailings are sometimes deposited in impoundments with a natural soil foundation, such as bentonite clay, that has a very low permeability and can serve as a natural liner. In other cases, synthetic and natural materials are used in combination to form the impoundment's liner. However, there are no uniform standards, and in some cases tailings are still stored, even today, in impoundments that have no impermeable liner — either natural or synthetic.

Tailings impoundment systems, even those built with impermeable liners and state of the art dams, unfortunately are not foolproof. Impoundments require proper design, careful installation, diligent maintenance, and persistent regulatory inspection. At some sites, overloaded or unstable waste material has placed stress on liners, causing breaks and tears. In other cases, failures have occurred from manufacturing defects in synthetic material. Mishandling of the liner material during installation — for example, operation of equipment directly on the surface — has caused punctures and tears.

PHOTO: PHILIP M. HOCKER/ MINERAL POLICY CENTER

The longest of these haul trucks can transport up to 190 tons of earth. Many trucks operate 24 hours a day, seven days a week, carrying ore from a mine's pit to a processing mill or waste rock piles.

Since synthetic liners are bonded together from separate panels, leaks can occur in the bonded seams. Improper installation of the liner in winter on frozen ground can cause the liner to tear due to freeze and thaw shifts in the ground foundation. Repeated freezing, thawing, and drying can also cause low-permeability soils used in clay-lined impoundments to crack.[68]

Impoundment embankments can fail if improperly designed or located on an unstable foundation. Constructing the embankments with highly permeable or unstable material can lead to leaks or total collapse. Overfilling the impoundment can also overwhelm the embankment's stability. Likewise, impoundment designs that fail to provide adequate holding capacity for rainwater and snowmelt drainage, as well as the tailings, can lead to environmental catastrophes. Tailings impoundments and dams which are unable to contain or withstand heavy rains can allow tailings to flow over the top of the embankment, leach through the dam, or result in a total dam collapse.

Given all of these potential problems, it is not surprising that tailings impoundments frequently leak. U.S. Environmental Protection Agency monitoring of tailings and other waste facilities at eight active mine sites in the gold, copper, lead, phosphate rock, and uranium industries showed evidence of leakage of contaminants into groundwater and surface waters at almost all facilities.[69] Several examples are examined in Chapter 3.

Waste Rock Piles

Unlike tailings impoundments, waste rock piles are fairly simple structures. Material that does not contain the target mineral is removed from the surface or underground mine and deposited in large piles. The piles are sited as close to the mine as possible to minimize haulage costs. Because waste rock piles can be massive in size, and often contain acid-forming rocks and metal contaminants, they pose environmental hazards when exposed to wind and water. If exposed to natural elements, the sediments and pollutants in the waste rock can be transported off the mining site easily by wind, rainfall, snowmelt, and stream water drainage. Moreover, since most waste rock piles are not lined underneath, contaminants in the waste rock can seep into the groundwater below.

Effective waste rock management techniques isolate these piles from the natural elements and reclaim them after mining. These practices may include placing the piles away from the influence of natural surface water drainages. In addition, operators sometimes divert surface water drainage and streams around the waste piles, or install drains that are intended to pass water through or around the piles. These diversion techniques minimize the amount of water that comes directly in contact with the waste material. Some operations further isolate waste piles from the elements by "capping" them — covering them with low-permeability materials like clay to prevent water and air from infiltrating — and by reclaiming them with topsoil and vegetation. In rare cases, waste rock piles are placed on an impermeable, synthetic liner which protects groundwater from being contaminated by pollutants that leach from the waste rock.

Unfortunately, many billions of tons of waste rock left on the American landscape have no protection at all against wind, water, rain, or snow. This is, in part, the legacy of over a century of accumulated mining wastes generated when mine waste disposal was given almost no thought whatever. Even today, however, when environmental management techniques exist, a significant portion of tailings and waste rock is still deposited directly on the ground without liners or other protective barriers.

Phosphate Mining Waste

While waste disposal methods and environmental problems from most metal and non-metal mining are similar, waste generated from phosphate mining deserves special attention.

A grey non-metal, phosphate rock is mined mostly to produce fertilizers, pharmaceuticals, and animal feed. Phosphate rock, which originated from the bone and shell remains of tiny sea organisms, contains phosphorus, an essential plant nutrient. Raw phosphate rock is usually converted by processing techniques into phosphoric acid, a clear colorless liquid used in a variety of end products. At some phosphoric acid plants, radioactive uranium is produced as a byproduct, since phosphate rock deposits naturally contain uranium in trace amounts.

In the hardrock mining industry, phosphate rock mining generates waste quantities exceeded only by copper mining. Phosphate accounts for approximately one quarter of the tailings and waste rock generated by hardrock mining.[70] In 1987, phosphate mining produced 408 million tons of waste.[71]

Phosphate mining makes intensive use of water. First, the overlying material, called overburden, is removed by dragline (a large earth-moving machine) or excavator to expose the phosphate ore, which is generally formed in a horizontal seam consisting of sand, clay, and phosphate rock. High pressure hoses then direct a torrent of water against the underlying layer of ore. The resulting water/rock slurry is collected and transported to a mill where it undergoes beneficiation. In the mill, phosphate rock is graded by selective sizing with mesh screens. Mills also employ a flotation process to separate phosphate rock that is intermingled with fine sand particles. Large volumes of water-laden clays known as clay slimes, usually impounded behind dams at the mine site, are the largest waste byproduct of this process. When phosphate rock is treated with sulfuric acid to produce phosphoric acid, a white, silty waste product called phosphogypsum is created. Phosphogypsum and highly acidic processing water are pumped as a slurry into tall waste stacks, known as gypsum stacks, and remain permanent features of the local landscape.

Phosphate mining districts are littered with sprawling clay slime ponds and gypsum stacks. Clay slime settling ponds, which are typically surrounded by compacted earth dams up to 60 feet high, store huge amounts of water, slowly releasing it into the groundwater over many years as the clay settles. Gypsum stacks can tower as high as 150 feet and cover over 400 acres. Gypsum stacks represent some of the highest structures in Florida (see Chapter 5, Photo).

Wastes generated by phosphate mining have caused harm to water resources. Phosphogypsum and processing water often contain elevated levels of uranium, radium, gross alpha particles (radioactive emissions), cadmium, arsenic, chromium, lead, sodium, fluoride, manganese, and iron. Many of these substances, which are natural components of phosphate rock ore, occur at toxic levels in the waste impoundments. On several occasions, processing water and gypsum stacks have contaminated underlying aquifers and drinking water wells.[72] Although clay slime ponds are not believed to pose a serious risk to groundwater quality,[73] they have on several occasions collapsed and released floods of sediment-laden wastewater, choking nearby rivers with sediment.

Solution Mining Waste

Because dump and heap leaching allow the economic extraction and processing of large-volume, low grade ores, the amount of waste generated by mining has naturally skyrocketed. The amount of material handled at U.S. gold mines (a good reflection of waste tonnage) increased from 164 million tons in 1984 to 597 million tons in 1992. Total materials handled at copper mines increased from 505 million tons to 808 million tons between those same years.[74,75] These growing waste volumes pose an expanding management problem and an increased risk to water resources.

After several leaching cycles, many leach heaps and dumps are now approaching the end of their useful life, since the ore has given up most of its mineral values. Now these piles will become part of mining's daunting waste problem. Spent leaching material, like most other mining wastes, cannot be easily removed or relocated. These sprawling and towering deposits of spent ore store huge quantities of cyanide, acid-forming sulfides, and heavy metals. If not properly managed, reclaimed, and monitored, these waste deposits will be a persistent threat to water quality.

Although cyanide often breaks down rapidly in the open-air environment and does not bioaccumulate in organisms, it is still an acutely toxic substance that requires careful management (see Appendix). While most modern mines place some sort of synthetic liner under leaching pads to prevent seepage, poor management and operator errors have caused breaks in liners and leaks of cyanide solution, threatening the environment and public health. Such accidents have happened with alarming frequency. Similarly, copper leaching operations present serious risks to water quality. The sulfuric acid solutions that are used in copper leaching and the heavy metals contained in the waste dumps can infiltrate groundwater or wash into surface streams.

Because there is no consistent Federal requirement for reclamation of these heaps, once they are exhausted they remain as permanent fixtures on the landscape, with their pollutants exposed to the environment. Unless adequate reclamation steps are taken, such as rinsing out and treating the heap's contaminants and capping the heap with material that is highly impermeable, water pollution problems can continue at these mines long after mining has ceased.

On first appraisal, in-situ mining might appear to reduce mining's waste generation problem, because this method does not generate waste rock or tailings. Despite its seeming environmental advantages, however, in-situ mining poses grave risks to groundwater quality. When they close, in-situ operations leave behind liquid wastes that may contain dissolved metals at harmful concentrations. Leaching solutions, besides freeing the target metal, also mobilize other metals, potentially raising their concentration in groundwater to unsafe levels. Unless carefully controlled, the solutions can migrate outside the immediate mining area and contaminate nearby aquifers.

Wasteful Past, Wasteful Future

As mining has become more mechanized and efficient over the years, hardrock mines have been able to handle more rock and ore material than ever before. Consequently, mine waste generation has intensified. Newer mining techniques such as cyanide heap leaching and SX/EW, which make it profitable to mine low grade ore, will accelerate waste generation in the future. Reliance on decreasing grades of ore for many metals will reinforce this trend, because more ore must be processed to obtain the same amount of metal.

This reality and the environmental threat to water resources it poses create an increasing need for the mining industry to adopt and consistently apply practices that minimize the impacts of waste. Unfortunately, the industry has failed to adopt such practices, and the agencies charged with regulating the industry's practices have not properly enforced environmental standards adequate to protect water quality. The result has been an ever-increasing amount of unmanaged mining waste and the destruction or imperilment of many of our nation's critical streams and groundwater supplies.

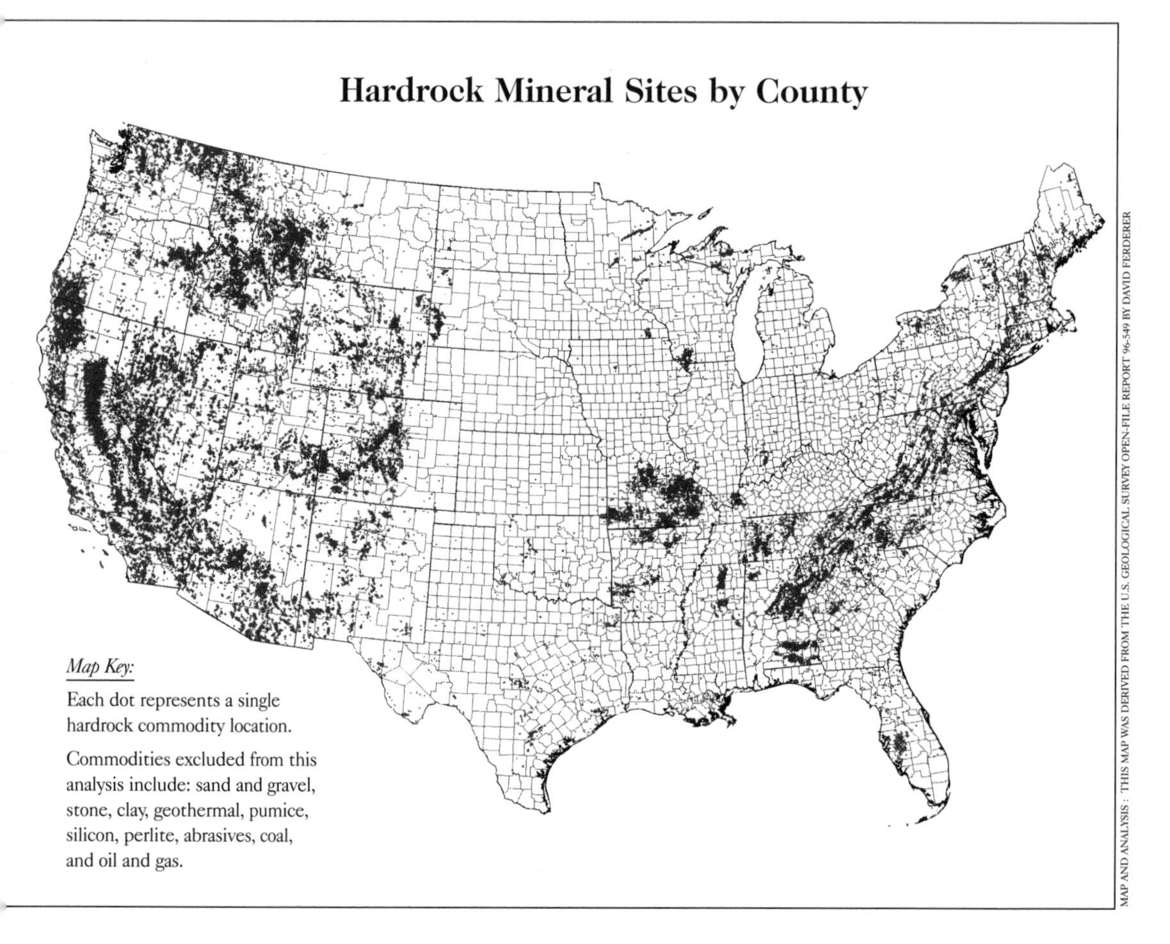

This map, constructed using Geographic Information Systems (GIS), shows the distribution of approximately 106,000 hardrock and metal commodity locations in the lower 48 states, excluding industrial and energy-related sites. According to the U.S. Geological Survey, these sites represent 52 percent of all mineral locations in the former U.S. Bureau of Mines Minerals Availability System (MAS) database.

♦ ♦ ♦

Notes

1. Telephone communication with Fred Sanchez, Human Resources Department, Pinto Valley Mine Division, Magma Copper Corp. (21 November 1995).

2. Telephone communication with Dan Gaynor, Mannesmann Demag Corporation (23 October 1995).

3. MAGMA COPPER COMPANY, 1994 ANNUAL REPORT, p.9 (1995).

4. In 1995, 209 million tons of municipal solid waste were generated in the United States. U.S. ENVIRONMENTAL PROTECTION AGENCY, OFFICE OF SOLID WASTE, MATERIALS GENERATED IN THE MUNICIPAL SOLID WASTE STREAM, p.26 (1995).

5. Letter from Randall W. Thomas, Director of Planning, Pinto Valley Mining Division, Magma Copper Company, to Carlos Da Rosa, Mineral Policy Center, 27 November 1995. A football field has an area of approximately 1.32 acres.

6. Bob Golfen, *Stream Fouled by Mine Runoff,* THE ARIZONA REPUBLIC, 7 February 1993, p.B1.

7. U.S. ENVIRONMENTAL PROTECTION AGENCY, REPORT TO CONGRESS: WASTES FROM THE EXTRACTION AND BENEFICIATION OF METALLIC ORES, PHOSPHATE ROCK, ASBESTOS, OVERBURDEN FROM URANIUM MINING, AND OIL SHALE (31 December 1985) [Hereinafter REPORT TO CONGRESS].

8. REX BOSSON & BENSION VARON, THE MINING INDUSTRY AND THE DEVELOPING COUNTRIES (Washington, D.C.: U.S. Government Printing Office, 1978), *cited in* John E. Young, *Mining the Earth,* WORLDWATCH PAPER 109, p.21 (July 1992).

9. HOWARD N. & LUCILLE N. SLOANE, PICTORIAL HISTORY OF AMERICAN MINING, p.1 (1970).

10. T.A. RICKARD, A HISTORY OF AMERICAN MINING, pp.24-25 (1932).

11. *Id.,* pp.147-152.

12. *Id.,* pp.6-7.

13. SLOANE, *supra* note 9, p.58.

14. FLORIDA PHOSPHATE COUNCIL, BILLIONS OF YEARS AGO (undated).

15. Waverly B. Lowell, *Where Have All the Flowers Gone? Early Environmental Litigation,* PROLOGUE, p.251 (National Archives and Records Administration, Fall, 1989).

16. Marilyn Ziebarth, *California's First Environmental Battle,* CALIFORNIA HISTORY, p.276, (Fall 1984).

17. *Id.*

18. *Id.*

19. Lowell, *supra* note 15, p.251.

NOTES

20. 18 F. 753, 756 (C.C.A. 1884).

21. Ziebarth, *supra* note 16, pp.276-279.

22. Ziebarth, *supra* note 16, p.276.

23. Charles V. Averill, *Placer Mining for Gold in California,* CALIFORNIA DIVISION OF MINES BULLETIN 135, p.144 (October, 1946).

24. RICKARD, *supra* note 10, p.91.

25. SLOANE, *supra* note 9, p.175.

26. U.S. ENVIRONMENTAL PROTECTION AGENCY, MINING WASTE NPL SITE SUMMARY REPORT, CALIFORNIA GULCH, LEADVILLE, COLORADO, p.3 (21 June 1991).

27. David *Fanning, Bleeding From a Million Wounds,* HIGH COUNTRY NEWS, 20 November 1989, p.11.

28. Edwin Dobb, *Pennies from Hell,* HARPER'S MAGAZINE, October, 1996, p.40.

29. U.S. ENVIRONMENTAL PROTECTION AGENCY, MINING WASTE NPL SITE SUMMARY REPORT, TORCH LAKE, HOUGHTON COUNTY, MICHIGAN, p.1 (21 June 1991).

30. U.S. ENVIRONMENTAL PROTECTION AGENCY, MINING WASTE NPL SITE SUMMARY REPORT, CARSON RIVER, LYON AND CHURCHILL COUNTIES, NEVADA, p.1 (21 June 1991).

31. U.S. ENVIRONMENTAL PROTECTION AGENCY, MINING WASTE NPL SITE SUMMARY REPORT, SILVER BOW CREEK/BUTTE AREA SITE, BUTTE, MONTANA, p.5 (21 June 1991).

32. MINING WASTE NPL SUMMARY REPORT, TORCH LAKE, HOUGHTON COUNTY, MICHIGAN, *supra* note 29, p.1.

33. TENNESSEE VALLEY AUTHORITY, A PLAN FOR REVEGETATION COMPLETION OF TENNESSEE'S COPPER BASIN, TVA/ONRED/LER-86/51, p.8 (June 1986).

34. U.S. ENVIRONMENTAL PROTECTION AGENCY, MINING SITES ON THE NPL (August, 1995).

35. U.S. BUREAU OF MINES, MINERALS YEARBOOK 1992 (VOLUME I), p.84 (1993).

36. *Id.* (1993).

37. Telephone communication with Dennis S. Kostick, U.S. Bureau of Mines (26 September 1995).

38. NEW STRATEGIES FOR GROUNDWATER PROTECTION (Eric P. Jorgensen, ed., Island Press 1989), *cited in* Young, *supra* note 8.

39. MDL INFORMATION SYSTEMS, INC., INFORMATION SHEET (1996).

40. REPORT TO CONGRESS, p.ES-5.

41. U.S. BUREAU OF MINES, MINERALS YEARBOOK 1992 (VOLUME I), p.82 (1993).

Notes

42. U.S. Environmental Protection Agency, Report to Congress on Special Wastes from Mineral Processing, pp.6-4,10-3,14-2 (July, 1990).

43. U.S. Bureau of Mines, Gold and Silver Leaching Practices in the United States, IC 8969, p.8 (1984).

44. U.S. Environmental Protection Agency, Office of Pollution Prevention and Toxics, RM2 Briefing on Cyanidation Mining, pp.9-10 (22 November 1991).

45. Telephone communication with John Lucas, U.S. Bureau of Mines (20 March 1995).

46. Magma Copper Company, Source Materials (1989) (as communicated to Mineral Policy Center by Daniel Edelstein, Copper Commodity Specialist, U.S. Geological Survey, 27 January 1997).

47. U.S. Environmental Protection Agency, Office of Solid Waste, Technical Resource Document: Copper, p.1-47 (August, 1994).

48. Telephone communication with Daniel Edelstein, U.S. Bureau of Mines (20 March 1995).

49. Daniel J. Millenacker, *In-Situ Mining,* Engineering and Mining Journal, pp.56-57 (1989).

50. Brown and Caldwell, Magma Florence In-Situ Project Aquifer Protection Permit Application (Volume II of V) (Prepared for Magma Copper Company), pp.1-4 (January, 1996).

51. In-Situ Project Manager, Asarco Santa Cruz, Inc., The Santa Cruz In-Situ Copper Mining Research Project (Information Pamphlet) (undated).

52. Peter G. Chamberlain & Michael J. Pojar, U.S. Bureau of Mines, Gold and Silver Leaching Practices in the United States, IC 8969, pp.34-35 (1984).

53. Telephone communication with Orville Kiehn, U.S. Environmental Protection Agency (13 January 1997).

54. Energy Information Administration, Uranium Industry Annual 1995, DOE/EIA-0478, p.ix (May, 1996).

55. *Id.*

56. Energy Information Administration, Decommissioning of U.S. Uranium Production Facilities, DOE-EIA-0592, p.30 (February, 1995).

57. *Id.*, p.29.

58. Report To Congress, p.2-11.

59. John L. Dobra and Paul R. Thomas, The U.S. Gold Industry 1992, p.16 (Nevada Bureau of Mines and Geology, Special Publication 14, 1992).

60. Based on the amount of gold in one average 14 carat gold man's wedding band.

NOTES

61. In the United States, in 1993, an estimated 2 million tons of copper were produced compared with approximately 360 tons of gold. U.S. BUREAU OF MINES, MINERAL COMMODITY SUMMARIES 1994, pp.54,73 (1994).

62. These approximate waste figures have been derived by adding the amount of mine waste (which consists mostly of waste rock) to the amount of tailings. The Bureau of Mines produces figures for the amount of mine waste but not tailings. An approximate figure for tailings has been obtained by subtracting BOM figures for mine production from BOM figures for amounts of raw ore. The BOM stated that its 1992 mine waste and raw ore figures for copper are subject to revision. U.S. BUREAU OF MINES, MINERALS YEARBOOK 1992 (VOLUME I), p.82 (1993).

63. OFFICE OF TECHNOLOGY ASSESSMENT, MANAGING INDUSTRIAL SOLID WASTES, p.30 (February, 1992).

64. REPORT TO CONGRESS, p.3-14.

65. H.E. Johnson and C. L. Schmidt, *Clark Fork Basin Project Status Report and Action Plan,* Clark Fork Basin Project, Office of the Governor; Helena MT, 1988, *cited in* Johnnie N. Moore and Samuel N. Luoma, *Hazardous wastes from large-scale metal extraction,"* ENVIRONMENTAL SCIENCE & TECHNOLOGY (Volume 24, No. 9) (1990).

66. ENVIRONMENTAL PROTECTION AGENCY, MONTANA SUPERFUND SITES STATUS REPORT (September, 1995).

67. UNITED STATES COMMITTEE ON LARGE DAMS, TAILING DAM INCIDENTS, p.2 (November, 1994).

68. MINE WASTE MANAGEMENT, pp.346-347 (Ian P.G. Hutchison and Richard D. Ellison eds., 1992).

69. REPORT TO CONGRESS, p.4-51.

70. OFFICE OF TECHNOLOGY ASSESSMENT, *supra* note 63, p.30.

71. *Id.*, p.29.

72. U.S. ENVIRONMENTAL PROTECTION AGENCY, OFFICE OF SOLID WASTE, REPORT TO CONGRESS ON SPECIAL WASTES FROM MINERAL PROCESSING, pp.12-20 to 12-25 (July 1990).

73. U.S. ENVIRONMENTAL PROTECTION AGENCY, OFFICE OF SOLID WASTE, TECHNICAL RESOURCE DOCUMENT, EXTRACTION AND BENEFICIATION OF ORES AND MINERALS, PHOSPHATE AND MOLYBDENUM, p.23 (November, 1994).

74. U.S. BUREAU OF MINES, MINERALS YEARBOOK 1984 (VOLUME I), p.22 (1985).

75. U.S. BUREAU OF MINES, MINERALS YEARBOOK 1992 (VOLUME I), p.82 (1993).

♦ ♦ ♦

Mining's Impacts on Water, Wildlife, and People

3

In the early 1900s, mining companies dug the Yak Tunnel four miles into Iron Hill to drain water out of the silver, lead, and zinc mines of Leadville, Colorado.[1] *There has been no mining at Iron Hill for decades — but the tunnel still spews orange, acidic, metal-laden water from the mines into California Gulch. The polluted water then flows into the Arkansas River. Each year, the Yak Tunnel discharges 210 tons of heavy metals — such as cadmium, lead, copper, manganese, iron, and zinc — into California Gulch. Picking up additional contaminants as it flows through mounds of waste rock and tailings, the water of California Gulch ultimately contributes more than 80 percent of the heavy-metal contamination that poisons the Arkansas River.*[2] *California Gulch is now a Superfund National Priority List site for toxic-waste cleanup.*

Wildlife surveys conducted by the State of Colorado have found a complete absence of fish in California Gulch and the upper Arkansas river just below the Gulch.[3] *Other studies have shown that brook trout found farther down the Arkansas River below California Gulch have accumulated high levels of copper and zinc in their livers.*[4]

Mining also threatens human health in the Leadville District. The EPA has found that California Gulch contaminates nearby groundwater, which contains concentrations of heavy metals in excess of national primary and secondary drinking water standards.[5] *For many years this groundwater was used as a source of local drinking water. Now, after testing, many of the drinking-water wells which had drawn up this polluted groundwater have been abandoned because of health risks.*[6]

Decades of television and newspaper stories have conditioned us so that when we think of water pollution we tend to think of leaking steel drums of man-made chemicals, or overturned tank trucks. Those sources exist at mines, too, but mining creates its own, much larger and more persistent, forms of water pollution — pollution based on the mined ores and waste minerals themselves. Acid mine drainage, toxic heavy metals, and processing chemicals from mining wastes all can contaminate water. These mining contaminants kill aquatic organisms, or in lower concentrations can interfere with the organisms' growth and reproduction. Sediments from eroding waste piles choke streams, destroying aquatic habitat and increasing chances of flooding. Mining pollutants also pose a risk to human health.

In addition, the use and waste of water at all stages of the mining process make it unusable for wildlife, agriculture, other industrial purposes, and as community drinking water supplies — a particular problem in arid parts of the West where water resources are very limited.

Leadville, Colorado, is but one example of the nationwide array of harms or threats to water resources from mining. The U.S. Geological Survey has identified 48 watersheds around the country, each of which contains at least 200 operating or historic hardrock mine sites (see Photo Insert).[7] These mining operations, past and present, have created hundreds of documented cases of wildlife casualties, streams and rivers without life, empty or poisoned wells, and fouled land. The following cases are only a few examples:

- The Coeur d'Alene Basin in northern Idaho has been contaminated by nearly a century of lead, silver, and zinc mining. In all, mines and smelters discharged over 70 million tons of tailings and slag into the Coeur d'Alene River and its tributaries. The 21-square-mile Bunker Hill Superfund site, which straddles the South Fork of the Coeur d'Alene River at Kellogg, Idaho, is one of the nation's largest Superfund sites. The mining pollution at this site is a major health threat to residents in the region, and it has devastated fish, migratory bird, and other wildlife populations. The victims include scores of tundra swans found dead in recent years with elevated lead levels in their tissues.[8] In February 1996, during one flood alone, more than one million pounds of lead from mining wastes washed down the Coeur d'Alene River into Lake Coeur d'Alene.[9] Mining wastes cover the lake bottom, as deep as 15 inches in some places.[10] In 1996, the U.S. Justice Department filed a lawsuit against eight mining companies in an effort to hold them financially responsible for cleanup of the massive environmental damage their works have caused in the region.[11]

- Acid mine drainage from waste rock and tailings at the Molycorp molybdenum mine in northern New Mexico are the primary cause of the decimation of a once-thriving trout population along the middle section of the Red River.[12] Local farmers believe that spilled tailings from the mine have also contaminated *acequias* (water ditches) long relied upon by area residents to irrigate their crops.[13]

- Between 1983 and 1992, at least 1,018 birds were killed when they drank cyanide-poisoned water from heap leach solution ponds at the Wharf Gold Mine complex in the Black Hills of South Dakota. State

Colorado Streams Affected by Metals

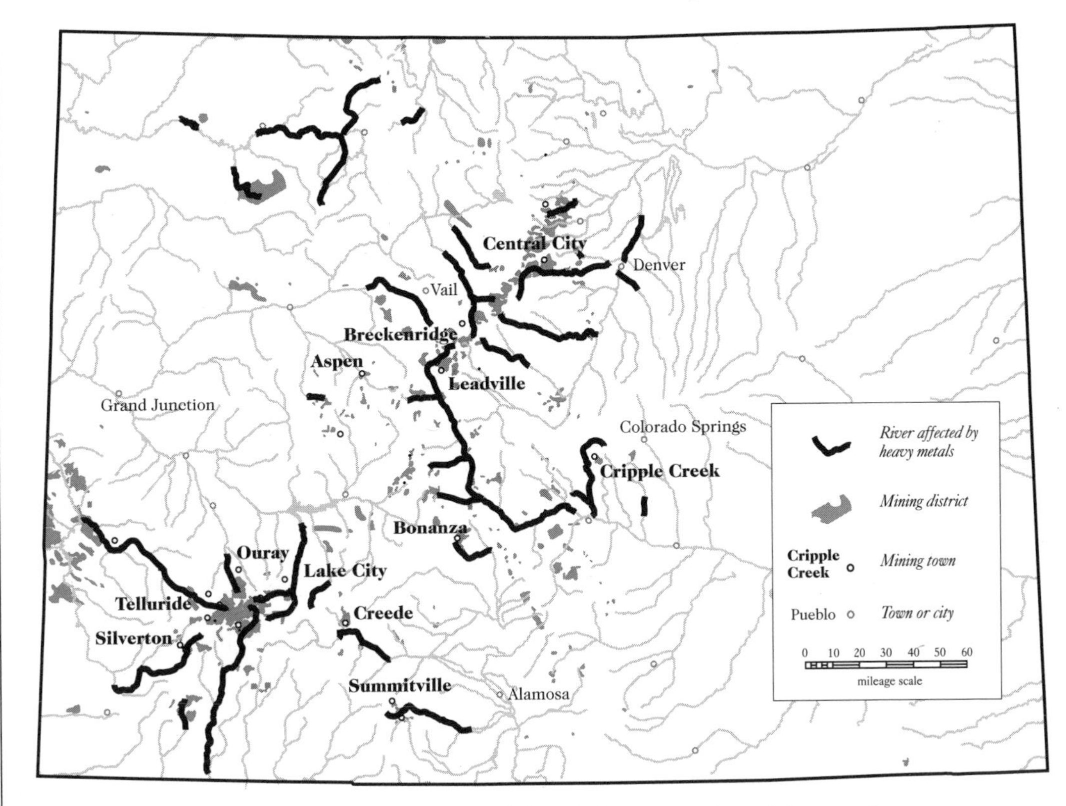

FIGURE MODIFIED FROM U.S. GEOLOGICAL SURVEY OPEN-FILE REPORT 94-264, P.95, "AFFECTED STREAMS FROM COLORADO WATER QUALITY CONTROL DIVISION," 1989 COLORADO NONPOINT ASSESSMENT REPORT.

Mining districts throughout west central Colorado have contaminated rivers and streams.

wildlife officials report that this massive wildlife kill encompassed 47 different species.[14] The wildlife death toll continued to climb in 1995 when heap leach solution ponds overtopped after heavy rains. The overflow carried mine wastewater containing ammonia and cyanide into a tributary of Spearfish Creek, causing the death of more than 300 fish.[15]

- Mining's water consumption, as well as mining's contamination, can cause severe impacts. Groundwater pumping by Nevada gold mines has lowered local water tables as much as 1,000 feet, causing natural springs to dry up.[16]

The environmental impacts of hardrock mining go far beyond the boundaries of a particular mining site. Transported by water, mining's pollutants can spread hundreds of miles from the source of contamination. While a wasteful and polluting mining operation prospers, wildlife and humans downstream become its victims. When mines close, the problems they leave behind frequently become the concern and the burden of someone else — a local community and, often, the U.S. taxpayers.

How Mining Contaminates Water

Mining contaminates water in four primary ways. A massive threat comes from acid mine drainage, acid runoff that results when iron sulfide minerals and other sulfides found in ore and mine waste rock are exposed to air and water. This drainage running off the mine site makes once-clean water in its path unfit for life.

A second source of contamination is toxic heavy metals such as cadmium, copper, zinc, and lead. Prior to mining, these metals are locked in rock underground. When the rock is unearthed, the metals can be set free by wind and water drainage and can spread through the environment in dangerous concentrations. Wildlife can tolerate these metals even less than humans can.

Chemical agents used to separate and recover minerals from ore provide a third source of contamination. This contamination can be very long-lasting. One of these chemicals, mercury, is no longer used in mining ore-processing in the United States, but it continues to affect the streams into which it was once dumped, in some cases more than 150 years ago. Today, such chemicals as cyanide — a mineral processing agent widely used in gold mining — pose a new chemical threat to wildlife and humans.

The final major category of water pollution from mining is erosion and sedimentation. When wind blows across a mine, and water (rainwater, snowmelt, surface streams) drains from a mine site, these natural forces can erode mine waste material and carry the sediments off the site, where they can clog river beds, smother aquatic life and habitat, and kill vegetation.

Acid Mine Drainage

Mining unearths mineral compounds and metals that were deposited in place in an oxygen-free environment and have been buried for millions of years. As Tom Aley and Wilgus Creath state in Chapter 5, "The opening of a mine, whether that mine be surface or underground, is similar in many ways to digging into a sealed toxic waste dump." Acid mine drainage occurs when iron sulfide minerals, such as iron pyrite (known as "fool's gold") and pyrrhotite, are present in the ore. When exposed, these iron sulfides react with water and oxygen to create sulfuric acid — H_2SO_4. (Other metal sulfides can also generate acid, but iron sulfides generate the greatest amount of acid mine drainage.) Sulfuric acid is the same compound that produces acid rain. Iron sulfides are common components of metal ores.[17] Acid mine drainage is often 20 to 300 times more acidic than acid rain.[18] Many acid mine drainage impacted streams have a pH value of 4 or lower — highly acidic (see Chapter 1, Science Notes). Some acid mine drainage has even been measured with a pH below zero.[19]

Plants, animals, and fish are unlikely to survive in streams with a pH below 4. Even more mild levels of acidity can cause adverse impacts on the functioning and reproduction of aquatic species. Nature can usually adjust for small changes in the acidity of streams and lakes. Alkaline or basic materials in the soil, such as limestone and other carbonates, can neutralize runoff, keeping the pH within a range that can sustain life. Acid mine drainage, however, easily overwhelms nature's defenses.

The chemical reactions that produce acid mine drainage can take place even where there is no mining, when natural weathering processes expose sulfide rocks to the air and rain. However, mining dramatically accelerates the process of acid generation because it breaks open and moves entire mountains full of sulfide rock in a short time. Waste rock containing iron sulfides is a principal source of this contamination, particularly if the waste is deposited where water flows. Finely-ground sulfide-bearing tailings left over from the ore beneficiation process can also be a major source of acid mine drainage. Underground mines puncture ore bodies with adits, mine tunnels, and shafts that allow air and water to enter and react with sulfide materials that are exposed inside the mine. Acid mine drainage can leach out of the underground mine openings into streams and aquifers. In open-pit mines, "pit-wall reactions," where the sulfide minerals on the exposed sides of the pit excavation are moistened by precipitation or by groundwater seeps, can generate intense acid mine drainage flows.

The rate of acid production from inorganic oxidation of iron sulfides can be exponentially escalated by certain types of bacteria that thrive in an acidic environment. Native to many areas of the Earth, acidophilic bacteria (acid-loving), most prominently the species *Thiobacillus Ferrooxidans,* are triggered into action when acidity falls below a particular level. These bacteria catalyze the oxidation of iron, and the work of the bacteria can increase the production of acid from mine waste three- to six-fold.[20] These bacteria obtain energy, which they use for cell maintenance and growth, from the oxidation of sulfides. The authors of a 1988 study of mine wastes in California, commissioned by the California State Legislature, have observed that the combination of acid-enhancing bacteria and huge waste volumes at mine sites makes acid mine drainage especially difficult to control: *"Once the problem has started, it tends to get worse."* [21]

Acid mine drainage is not merely an intense problem; it is also extremely persistent. Acid mine drainage may be generated when a mining operation begins construction or enters its operational phase, but it may not appear then. In some cases, acid drainage may emerge years later, or only after a mine has closed — but it may then persist for decades, even centuries.[22] Once the key factors which cause acid mine drainage are in place at a site, acid will continue to be generated until all the iron sulfides are leached from the mine waste material or until steps are taken to completely seal off the sulfide rock source from oxygen and water.

Acid mine drainage is usually first recognized when someone notices that some streams or ponds at a mine site are filled with orange water. Acid waters dissolve and mobilize many kinds of metals, such as iron, copper, aluminum, cadmium, and lead. These, particularly the iron, precipitate out as the water becomes less acidic, and coat stream bottoms with an orange-, red-, or brown-colored slime called *yellowboy.*

Waters flowing into Fisher Creek cover the stream bottom with an orange and brown slime. The acidity, caused by pyrite rocks, has overwhelmed repeated efforts by the Forest Service to revegetate spoiled lands in this Montana mining district (see also Chapter 8, Photo).

PHOTO: PHILIP M. HOCKER/ MINERAL POLICY CENTER

Today, these effects can be seen at an abandoned underground gold mine in Montana, where orange, acid waters flow from a mine drainage tunnel into Fisher Creek (shown above), located only a few miles from the border of Yellowstone National Park. As a result of the steady leakage of pollutants, the stream is severely contaminated: It has no fish or plant life. The stream bottom is coated with bright orange yellowboy slime. Although the tunnel contributes only 3 percent to the creek's water volume during low flow conditions (most of the year) it is responsible for 80 percent of the creek's heavy metal load during these times.[23, 24] A few miles downstream, Fisher Creek carries its contamination into the Clarks Fork River, Wyoming's only Wild and Scenic River and a premier trout stream.

Mining companies and water-quality regulators try to predict in advance whether the runoff from a mine or waste dump will be acidic, in order to plan control measures to stop acid pollution. However, it is very hard to make an accurate prediction of acid mine drainage prior to mining. Determining the potential for acid mine drainage depends on the evaluation of a multitude of factors that are particular to a mine site — for example, rock chemistry, the projected exposure of specific rock surfaces to air and water, the presence of any natural, acid-neutralizing materials in the rock, where mining wastes are to be placed, climate and rainfall data, surface and underground water drainage patterns, and the prevalence of acid-generating bacteria.

The most common procedure used to predict acid mine drainage is known as *acid-base accounting*. It works by measuring the bulk amounts of acid-generating and acid-neutralizing material in samples of material. Minerals containing sulfur, particularly sulfides such as pyrite, have the potential to generate acidity when exposed to water and oxygen; on the other hand, other groups of minerals such as carbonates like calcium carbonate ($CaC0_3$, the chief component of limestone) are effective in buffering or neutralizing acidity.

In acid-base accounting, the amount of acid-forming minerals present in a sample is measured to determine that sample's potential to produce acidity. A similar evaluation of the alkaline material in the sample is used to determine its ability to neutralize acidity. The acid-generating and acid-neutralizing potentials of the sample are expressed as numerical values. These values are then compared to predict the likelihood that the waste will generate acid mine drainage.

However, acid-base accounting has an important shortcoming: It often fails to predict acid mine drainage accurately because it fails to take into account other environmental and geochemical factors that affect whether a given material or mine will produce acid mine drainage. For example, acid-base accounting does not factor in the previously discussed pivotal role that bacteria can play in dramatically increasing acid production. The specific crystalline structure of a mineral also affects how active the acidic and alkaline compounds within the material will be.

Some mine companies and environmental consultants use a more advanced method, *kinetic testing*, as a supplement to acid-base accounting. Kinetic testing attempts to reproduce more of the factors that affect the rate of generation of acid mine drainage from a given material in a real-world setting. In a typical kinetic testing method, a sample of mine waste is placed in a cylindrical chamber or other closed container. In order to simulate the waste sample's natural environmental setting, such factors as air, water, and bacteria can be introduced into the chamber. The impact of these factors on acid generation in the waste sample can then be measured over time. The sample can also be tested for other characteristics such as concentrations of dissolved metals. In contrast to acid-base accounting, kinetic tests on mine wastes use a larger sample volume and are run for a longer period of time — often months.

Kinetic testing, properly carried out, is a much more sophisticated and reliable method of acid-production prediction than acid-base accounting. Even so, while kinetic tests can provide a good indication of whether mining wastes are likely to generate acid, they cannot be relied upon to precisely predict the future quality of mine drainage (such as pH levels and metal concentrations). Moreover, good kinetic testing takes a considerable time — just when mining companies are in a hurry to start digging. There is a lot of pressure to rush to a quick evaluation — and a quick evaluation is more likely to underestimate the level of acidity than to overestimate it. Thus, even kinetic acid-drainage testing methods have not been infallible in practice. [25]

- The Rain Gold Mine, near Elko, Nevada, has been combating a serious acid mine drainage problem since 1990, when surface water drainage from the mine's waste rock piles began generating acid, contaminating two miles of nearby Dixie Creek. The drainage also contained elevated levels of mercury and arsenic. The owner, Newmont Mining, did not anticipate the acid mine drainage problem because acid-base accounting tests on rock samples indicated that the mine's wastes were not likely to generate acid.[26]

Toxic Metals

Depending on geological factors, the metals found in mining waste may include arsenic, cobalt, copper, cadmium, chromium, gold, iron, lead, silver, and zinc. Many of these metals are necessary in trace amounts to sustain some life forms in nature. In higher concentrations, however, these metals can be highly toxic. Exposure to these metals may result in fish and wildlife deaths, the deterioration of wildlife habitat, or the disruption of the natural food chain. In general, the tolerance threshold for heavy metals is much lower for wildlife than it is for humans. However, mining companies, and even government regulators, often rely on water quality standards which were established to protect human health only. While these drinking-water standards set allowable concentration limits on heavy-metals that are believed to be safe for humans, they can still prove deadly for some plants and wildlife.

Metals tend to dissolve and mobilize more easily in the acid waters produced by acid mine drainage. When acid waters pass through waste rock and tailings, metals are leached out and dissolve readily in the runoff. Carried in water, the metals can travel long distances, contaminating streams and groundwater. When metals are in dissolved form they are more readily absorbed and accumulated by plant and animal life, and are therefore generally more toxic than when precipitated out or solid. Because metals can bioaccumulate (be stored) in animal tissues and plants, they can be passed to other living things through the food chain. The coupling of acid mine drainage with high concentrations of heavy metals is hardrock mining's most serious threat to the environment.

- One of the worst examples of environmental damage from metal contamination is western Montana's Clark Fork River. Surrounded by

Science Notes

PPM and PPB

PPM and PPB

Concentrations of metals and other contaminants in liquids is defined by the amount of *solute* (a dissolved substance like metal) in a unit amount of *solvent* (a dissolving medium like water). The concentrations of mining contaminants in water are typically reported as milligrams per liter (mg/L) or micrograms per liter (μg/L). These concentrations can be stated equivalently as *parts per million (ppm)* and *parts per billion (ppb)*, respectively, to state the ratio of the mass of solute to the mass of solvent.

- For example, since a liter of water weighs one kilogram (kg) (i.e. a thousand grams), one milligram (mg) of lead in a liter of water (1 mg/L) is equivalent to a lead concentration of one part per million (1 ppm). (A milligram is one one-thousandth of a gram.) One microgram (μg) of lead in a liter of water is equivalent to a concentration of 1 part per billion (1 ppb), since a microgram is one-one millionth of a gram.

- There are 455 grams in one pound and 28 grams in one ounce. Therefore, 1 ppb of lead in water would be equivalent to 1 oz of lead in 62.5 million pounds of water. Such minute concentrations of heavy metals, in the ppb and ppm range, can be toxic to aquatic organisms, mammals and humans.

three mining-related Superfund sites, the river served for decades as the dumping-ground for billions of tons of mine tailings containing copper, zinc, arsenic, lead, and cadmium. The sheer volume of waste that has been dumped into the river channel now exacerbates periodic flooding. The contamination of the Clark Fork extends downstream for 120 miles. Today, during dry, hot periods, water evaporates from wet sediment along the Clark Fork and dissolved metals form salt crusts — for example, blue and green copper sulfates — on the river banks. During heavy rains or snowmelt, the metals are washed back and redissolved in the river, where they destroy and inhibit aquatic life.

Four dams on the Clark Fork have trapped huge quantities of metal waste. For example, the Milltown Reservoir, a Superfund site located 120 miles downstream from Butte and Anaconda, the most intensive areas of mining and smelting, has accumulated approximately 100 tons of cadmium, 1,600 tons each of arsenic and lead, 13,000 tons of copper, and 25,000 tons of zinc[27] (see Chapter 7, Secondary and Tertiary Contamination by River Transport). In mid-July 1989, the continued lethal power of these metals was demonstrated with chilling force when a major rainstorm caused the Clark Fork to overrun its banks, picking up a massive load of metal contaminants. The following day, thousands of dead fish, mostly trout, were found floating in the river. Similar fish kills occurred in 1981 and 1987.[28]

Chemical Processing Agents

Chemicals used in processing ore at mines form another category of mining-caused water contamination. A variety of industrial chemicals is used in different processes to assist the separation of the metal being sought in an ore body from the mass of other rock that surrounds it.

One of the first chemicals used to process mined ores was mercury. Mercury is a dangerous toxin, and it bioaccumulates through the food chain.

Since Roman times, miners have known that, if a mixture of powdered rock and gold or silver metal in water is washed over mercury, the gold and silver will adhere to the mercury (or "amalgamate" with it) while the worthless rock powder washes away. This useful property of mercury has been used to concentrate precious metals from the Classical era to the present day. In a typical mercury-gold amalgamation process, a sheet of copper metal is "dressed" with a thin coating of liquid mercury. A slurry of powdered gold ore is then washed over the plate. The gold particles adhere to the mercury-covered plate and form a mercury-gold amalgam. Later, the amalgam is heated to evaporate the mercury and recover the gold.

Mercury played an important role in the processing of gold and silver ores in the United States from Colonial times until about 1925. However, the mercury-amalgamation process usually only recovers about 60 percent of the gold in an ore. Chemists searched for many years for a way to recover a higher percentage of the total gold value, and eventually discovered the more-efficient cyanide process.

Mercury amalgamation is not used in the United States today because of its inefficiency, mercury's extreme dangers to human health, and the strict safety and environmental controls on its use. However, because it requires little capital investment in equipment, mercury amalgamation is still widespread in many parts of the world, such as South America and Africa. It is intensively used today in the Amazon River Valley, where a gold rush has attracted thousands of small, uncapitalized, independent miners (see Chapter 9, Amazon River).

Mercury was used by gold miners throughout America's 19th century gold rushes to extract fine gold particles from stream gravels. So great was the demand for mercury in those days that many mercury mines were built in California just to supply the needs of gold miners.

- Mercury also figured prominently in the processing of gold and silver from the rich Comstock lode, discovered near Virginia City, Nevada, in 1859. Mineral processing mills flushed rivers of mine tailings containing an estimated 7,500 tons of mercury directly on the banks of the Carson River.[29] More than a hundred years later, these uncontained wastes continue to wash mercury deposits into the Carson after rainstorms and snowmelts. Nevada authorities have declared health advisories against eating fish from the river, and a 50-mile stretch of it has been declared a Superfund site. So far, projected remedial costs for the first of two cleanup units is estimated to be $1 million.[30] Work has not yet begun on the second unit, a potentially much larger project.[31]

Today, cyanide, rather than mercury, is the main chemical agent used to process gold and silver ores. A workable process to concentrate gold and silver using cyanide was developed in Scotland in 1887; the new technique was immediately put to use in the newly-developed Witwatersrand gold fields in South Africa. A well-run cyanide-process gold mill typically recovers 97 percent of all the gold in the ores it treats.

While cyanide breaks down readily in the environment and does not bioaccumulate in organisms, it is still an acutely toxic chemical. Cyanide causes death among living organisms by blocking the transport of oxygen across cell walls. Fish are particularly sensitive to cyanide when it leaks into streams and rivers. Tiny amounts are also fatal to birds and mammals.

Water contamination from miners' cyanide occurs when protective liners beneath solution ponds at mines break or tear, when cyanide holding ponds overflow or breach after rainstorms, when piping systems carrying cyanide solutions fail, and when too much cyanide solution is used in leaching operations and discharged in tailings. There have been many recorded problems with cyanide:

- At the Echo Bay company's McCoy/Cove gold mine in Nevada's Carlin Trend, a series of eight cyanide leaks took place in 1989 and 1990, releasing a total of nearly 900 pounds of cyanide into the environment.[32]
- Pegasus Corporation's Zortman-Landusky mine in north-central Montana has experienced at least a dozen leaks and discharges of cyanide solution since 1982, resulting in contamination of streams and groundwater[33] (see Case Study in this chapter).

Across the United States, about two hundred million pounds of sodium cyanide solution are manufactured, shipped, and consumed in gold mining every year. A fraction of an ounce is lethal to humans, smaller amounts to wildlife.

Erosion and Sedimentation

Contamination from soil erosion and sediment pollution is another of mining's impacts to our water resources. Mining strips away vegetation, a natural anchor for rock and soil. Explosives and earthmoving machinery fracture and dislodge rock, much of which winds up in waste piles. Road building and the construction of mineral processing mills further disturb the land. Finely ground tailings are stored in impoundments. Ores are piled in hundred-foot-tall heaps for leaching with cyanide. All of this exposes more material to the elements, leaving that material prone to erosion — the transport of material over the earth's surface by natural forces, such as wind and water.

When erosion occurs, sediments are carried and transported into streams, lakes, and reservoirs where they accumulate. Even if the sediments themselves are not contaminated, they act as physical pollutants by changing the light, temperature, and oxygen conditions — conditions to which aquatic life has adapted over millions of years — in the waters that carry them. Some

PHOTO: HAGLER BAILLY SERVICES INC.

The Silver Bow Creek/Butte area is one of four separate, but contiguous Superfund sites located along the course of the Clark Fork River in Montana. Because there is no vegetation, mine tailings deposited on the banks of Silver Bow Creek are susceptible to wind and water erosion.

aquatic life may be disturbed or impaired; some destroyed. This may have dire consequences for the environment. Sediments can smother fish spawning grounds and, in extreme circumstances, the fish themselves. When sediments accumulate in streams and river channels, they reduce the capacity of waterways to carry stormwater runoff, which may cause flooding, causing even greater environmental impacts. Sedimentation also reduces the storage capacity of reservoirs. For all of these reasons, it is imperative that erosion control be a major part of every mining operation.

- Each day, the Grasberg Gold Mine in Irian Jaya, Indonesia, creates over 120,000 tons of mine tailings.[34] Since 1972, the U.S. based Freeport-McMoRan Copper & Gold Inc. has dumped mine tailings directly into the Aghawagon River. While the tailings are primarily non-toxic, the massive amount of material has clogged the river system and flooded more than 30 square kilometers of rainforest and agricultural lands[35] (see Chapter 9, Photo).

- The Ok Tedi copper mine in Papua New Guinea, a few hundred miles southeast of the Grasberg site, causes over one hundred million tons of tailings and sediment to wash down the Ok Tedi and Fly Rivers each year. Here, too, the mass of eroded and dumped rock is filling the natural river channel. If nothing is done, the mine's engineers predict that the mine's sediment will cause more than four hundred square kilometers of rainforest to be killed by flooding before the mine closes in 2010 (see Chapter 9, Ok Tedi Mine, Papua New Guinea).

Case Study

Zortman-Landusky Mine, Montana

Case Study: Zortman-Landusky Mine, Montana

The Zortman-Landusky gold mine sprawls 17,400 acres[36] in north-central Montana's Little Rocky Mountains. Since mining began in 1979, Zortman-Landusky's operations have been marred by a succession of leaks and discharges of cyanide and acid. These have resulted in cyanide contamination of streams and groundwater and a serious acid mine drainage problem. Despite the mine's lackluster environmental record, Montana state and federal agencies have only recently taken serious enforcement action against the mine.

Zortman-Landusky (Z-L) is owned by Zortman Mining Inc., a fully-owned subsidiary of Pegasus Gold Inc. It is located between the towns of Zortman and Landusky and borders the Ft. Belknap Indian Reservation — home to the Assiniboine and Gros Ventre Indian Tribes. These tribes consider the nearby Little Rocky Mountains sacred because they contain important cultural and spiritual sites.

Z-L was one of the first U.S. mines to extensively use gold heap-leaching technology, which involves spraying piles of ore with dilute cyanide solutions to extract microscopic flakes of gold. The mine consists of two open pits — the Zortman and Landusky pits, each a few miles apart, and 14 separate heap leach facilities.[37] Z-L is known for extracting the lowest grade of ore of any commercial gold mining operation in the United States — ultimately obtaining one ounce of gold for every 110,000 pounds of ore.[38]

The mine's troubles began early. In 1982, 780 gallons of cyanide-tainted solution leaked from a containment pond. Zortman Mining promised that such leaks would not be repeated. However, only a few months later, in October, a section of piping used in the mine's cyanide sprinkling system ruptured, releasing 52,000 gallons of cyanide solution onto lands and into creeks. A few days after the incident, a mine employee smelled cyanide — which has an almond smell — in his tap water. Testing revealed a cyanide concentration of 3.2 milligrams per liter (mg/L), well above drinking water standards, with cyanide levels in a local creek measuring as high as 22 mg/L.[39] These discoveries forced the shutdown of a local community water system. Despite this major release, cyanide leaks continued: over the next two years, eight separate cyanide spills occurred. Despite the scores of illegal releases, Pegasus paid only a $15,000 fine in connection with the October 1982 spill.[40]

Animal deaths provided the most dramatic evidence of cyanide contamination. In August 1983, two bighorn sheep were found dead on a cyanide heap leach pile. In 1991, 30 gulls landed on an unnetted cyanide pond and died.[41] That same year, a deer died after it sipped some cyanide solution that had trickled beneath a fence.[42]

Already concerned about Z-L disturbing their traditional sacred lands, the nearby Assinoboine and Gros Ventre Indian Tribes were also fearful about the prospect of the mine's contaminating their drinking water with heavy metals and cyanide. The Tribes' anxieties were exacerbated when a local physician tested blood of nine children on the Ft. Belknap Reservation

Zortman-Landusky Mine, Montana

and found elevated lead levels in six of them.[43] Although the tests did not prove a link between the elevated lead levels and Pegasus' mining, the result was cause for concern. The Tribes have more recently reported an increase in cancer and still births among members, though the lack of baseline data does not make it possible to prove a link between health problems and mining. Tribe members have long decried the insufficient monitoring of water quality in and around their reservation.

On one occasion, in September 1986, Z-L, without obtaining a permit, dumped 20 million gallons of treated cyanide solution onto 17 acres of land when a solution pond threatened to overflow after a heavy rainstorm.[44] The intentional discharge caused anxiety among residents and prompted the state to warn the mine that it could not discharge solutions without a water quality permit. Soon, trace levels of cyanide began to show up in Ruby Gulch, a small creek which runs off the mine site.[45]

Problems mounted as the mine demonstrated sloppy management of its heap leach operations. Several times over the next few years, leach pads suffered partial failures when they slipped under overloaded piles. According to state officials, one failure caused a large release of cyanide solution, creating a plume which contaminated groundwater for several years. In 1989, a Montana Department of State Lands Inspector found a leach pad overloaded by 75 feet — or approximately ten million tons — beyond the level allowed by the company's permit. Despite the risk of leakage or pad failure, the state did not take enforcement action or impose a fine.[46]

The mine also began to experience a serious acid mine drainage problem as it began to mine environmentally risky sulfide ores, which have the potential to generate acid mine drainage. Remarkably, since beginning operations, the mine had only received permits to mine less risky oxide ores. After a rainstorm in August 1993, an Environmental Protection Agency inspection confirmed that the mine had indeed been mining sulfide ores: several discharges from the mine contained elevated concentrations of heavy metals as well as high acidity. In fact, one stream had a pH of 3.0, more acidic than vinegar. The EPA then cited Zortman Mining for discharge of pollutants in seven unauthorized locations.[47] This federal action prompted Montana to sue Zortman Mining in August 1993 under the Montana Water Quality Act. The company was cited for illegal discharge of pollutants from eight locations at the mine.[48]

Montana engaged in fruitless negotiations with Pegasus and Zortman Mining for over a year. Finally, the EPA, at the state's request, sued the Zortman Mining Company and Pegasus Gold in June 1995 for violations of the Clean Water Act. The two Ft. Belknap Tribes also filed a similar suit against Zortman Mining and Pegasus Gold. The Tribes claimed that Z-L has damaged their historic water rights; they also demanded a better program of water quality monitoring on their reservation.[49]

On July 22, 1996, the EPA, state of Montana, Indian Tribes, and Pegasus Gold reached a $37.5 million agreement to settle their respective law suits. The settlement, the largest in Montana for violation of the state's water

Case Study

Zortman-Landusky Mine, Montana

quality standards, requires the mining company to pay fines to the state, federal, and tribal governments, and undertake water treatment, improved ground and surface water monitoring, and other environmental remediation work at the mine. The settlement requires Pegasus to conduct a study of water resources at the Fort Belknap reservation and to finance a health study of the reservation community.[50]

Z-L's serious environmental problems and the lawsuits against it have not disturbed the mine's ongoing plans to expand its operations into environmentally risky sulfide-rich ores. Z-L applied to permit its latest expansion in 1992. Known as the "Zortman Extension Project", the proposed expansion would increase the mine's size by one-third — its largest expansion to date. In part because of EPA insistence, the Bureau of Land Management (BLM) and Montana Department of State Lands have conducted a full-scale Environmental Impact Statement (EIS) on the expansion project, the first EIS submitted for any of the mine's 11 expansions. For its previous expansions, the mine never received more than cursory Environmental Assessments, which fall far short of the more thorough review provided in an EIS and do not provide for public comment. The final EIS for the mine was issued in March 1996. And in October 1996, BLM made the decision to approve Z-L's expansion, despite the mine's troubled environmental record.

PHOTO: JOHN SMART

The Zortman mine is situated directly above the town of Zortman, Montana. The polluted Ruby Gulch flows from the mine through the town.

Water Use, Water Waste

Mining does not merely pollute water that runs away from the mine site. Water used at the mine itself plays a pivotal role in all stages of mining, from excavation through processing. Water is used in drilling, dust control, dredging, and extraction; in washing, crushing, and grinding of ores; in flotation and leaching; in the transport of slurry; and in other operations.

Phosphate mining utilizes one-third of all water used in U.S. hardrock mining.[51] Most of this water is used for processing, but a substantial amount is also used to extract the phosphate rock itself. Modern phosphate mines recirculate much of their water, approximately 85 percent,[52] but because of loss through processing and evaporation, they must still replenish water with new sources.

While no single form of metal mining consumes as much water as phosphate mining, metal mining's water consumption rate is of great concern since a significant amount of that mining takes place in areas where groundwater is replenished slowly. Copper mining, for example, concentrated mostly in the arid West, uses approximately 200 gallons of water per ton of ore milled.[53] At this rate, the 299 million tons of copper ore milled in 1994 required 59.7 billion gallons of water,[54] a volume equivalent to 600 days — or nearly two years — of water use for a municipality the size of Austin, Texas.[55] Copper mines, too, recycle much of the water they use, but they must still draw on large quantities of new water to replace the water lost in processing and evaporation. In 1984, domestic copper operations recirculated approximately 60 percent of their water.[56] By the year 2000, copper mines are expected to recycle about 70 percent.[57] The amounts not recycled are still immense, and are growing because total mining activity is expanding.

But while miners need to obtain water to process their ores, getting rid of water in mines is even more urgent. All of the inventors of the steam engine, Savery, Newcomen, and James Watt, were striving to find a more efficient (and profitable) way to pump water out of underground coal mines. The Yak Tunnel cited at the beginning of this chapter was built to drain floodwaters out of the mines — letting gravity remove the water from the bottom instead of pumping it up with machines.

So today, companies often install groundwater pumping wells not only to obtain water for ore processing operations such as milling and concentrating, but also to keep underground and open-pit mines from flooding when mining operations reach down to below the water table.

Mines usually pump water by using a combination of in-pit wells and perimeter wells. In-pit wells, located inside open-pit mines, pump out water that has already entered the mine; perimeter wells, arranged at a distance outside and around the pit, intercept groundwater before it can seep into the pit. As the complex of wells pumps away groundwater, the water table lowers, forming what is called a "cone of depression." A three-dimensional drawing of the water table around the mine resembles a cone. This lowering of the water table decreases groundwater elevation miles away from the mining area. This lowering of the groundwater level can cause once-bountiful wells and springs

PHOTO: PHILIP M. HOCKER/ MINERAL POLICY CENTER

Despite the arid climate of the region, this vast Newmont/Barrick Goldstrike mining complex in the Carlin Trend of Nevada, consumes massive quantities of water for mining and milling.

to run dry, and can diminish the flow of rivers and streams, impacting nearby communities and wildlife habitat (see Chapter 5, Dewatering Impacts).

The impacts of mining's raw consumption of groundwater are particularly dramatic in northern Nevada, the locus of the nation's current gold mining boom. As of 1996, approximately 40 gold mining operations are located in this water-scarce area which is drained by the Humboldt River, Nevada's largest. Several gold mines in this area are pumping groundwater out of the underlying source at a rate of more than 10,000 gallons per minute. The Canadian-based Barrick Gold Corporation's Goldstrike mine withdraws water at a rate of more than 70,000 gallons per minute,[58] or 100 million gallons per day — about as much water as the entire city of Austin, Texas, uses daily.[59]

In a recent year, 1994, mining operations in the Humboldt River basin withdrew approximately 200,000 acre-feet of water,[60] more than enough water to supply all domestic water users in greater Seattle (population 1.1 million) over the same period.[61] The overwhelming bulk of these groundwater withdrawals was pumped at around 15 of the large-scale gold mines in the basin.[62] A significant portion of the water withdrawn — about 25,000 acre-feet per year — is consumed at the mines in mining and milling operations.

At current mining trends, 250,000 acre-feet of water will be consumed for mining and milling purposes in the basin over the next 10 years (1997-2007).[63] However, total water withdrawal and consumption for mining in the Humboldt River basin will likely rise above these current trends, because new mining projects are being planned and many existing operations are seeking to expand. In the spring of 1995, operating permits were being sought for the expansion or new opening of 13 different gold projects in the Humboldt River basin.[64]

Gold mines in the basin divert pumped water to irrigate fields, to create wetlands (in what is naturally desert), or to supply water to another user. Only a small fraction of this pumped water is restored to the aquifer whence it was pumped.

Where it can be established that pumping dries out a spring or well, a mining operation must, under Nevada law, replace the lost water, deepen the well that has been dried out, or reduce pumping so that prior water levels are restored.[65] Though this law seems to protect groundwater resources, Nevada has not yet officially recognized any instance of a mining company's diminishing another user's water right.[66] However, many water users, including ranchers and Native American communities, have complained about threatened or disrupted water supplies.[67] To protect themselves from legal liability for diminishing the water quantity of others, mines are increasingly buying out private land owners or trading private land for public land to create buffer zones that encompass a mine's complete cone of depression. Sometimes land is purchased by mining companies to serve as a disposal area for pumped water, or sometimes the purchased land is leased to ranchers and farmers, who now have no claim to the (diminished) water rights.

The long-range impacts of water use by mining operations in the Carlin Trend are unknown. The Humboldt River is Nevada's only significant waterway. It is also critical habitat to many species of plant, fish, and other animal populations. As the population of Nevada grows, so too does human demand on the Humboldt's water.

Despite the importance of water supplies in the region, and the alarming level of dewatering caused by mining, there has been no major study examining how large-scale gold mining in the Nevada desert will likely affect groundwater and surface water resources in the future. In 1994, the Sierra Club commissioned a brief study surveying the cumulative hydrologic impacts of mining in the Humboldt River watershed. This report estimated that mining's current and projected groundwater pumping may result in a total deduction of more than one million acre-feet of water in the Humboldt River basin.[68] The report also notes that most of the aquifer damage may come not during the course of pumping during active mining, but afterwards, when mines close and pumping ceases. At that time, the open mining pits that will be left behind (none of the pits will be re-filled with rock or soil when mining ends) are expected to become huge water sinks. Because the bottoms of the pits will lie below the water table, they will collect groundwater and draw flows away from the Humboldt River.[69] This report concludes by calling for a more detailed study by an independent agency.

The U.S. Bureau of Land Management, Nevada Engineer's Office, and the U.S. Geological Survey's Water Division have issued only a cursory statement on mining's water use in the Humboldt River basin.[70] Several federal and state agencies have begun a detailed impact study. The study is funded largely by mining companies, including Barrick Gold. The objectivity of the study is suspect. Study results are not expected to be available until the year 2000 or later. Meanwhile, northern Nevada continues to draw the interest of even more gold mining operations. The rate of groundwater pumping is being allowed to increase still further, before the full impact of this massive water displacement is understood.

- Around Tucson, Arizona, groundwater is being pumped at a startling 2.5 times the recharge rate (an aquifer's natural ability to replenish its water level). Mining, the largest industrial user of water in the Tucson area, is combining with agricultural and surging urbanization to place increased stress upon the areas's limited water supplies. As of 1996, mining accounted for approximately 15 percent of groundwater used in the Tucson area.[71] The three largest copper mines in the Tucson area increased groundwater withdrawals from 32,400 acre-feet in 1990 to 43,063 acre-feet in 1995.[72]

Impacts on Natural Life

Through all the ways mining damages our water supplies — water contamination with heavy metals, acid mine drainage, sedimentation, and water consumption and aquifer depletion — mining damages the myriad living things, including people, that depend on that water, as well.

While mining damages the natural environment in many ways, such as destroying wildlife habitat and surface vegetation, the most tragic environmental casualty of hardrock mining is the aquatic biosphere (see Chapter 6 for more detail). Acid mine drainage waters, toxic heavy metals, and sediments from mining damage the interlocking network of fish, insect, and microbes and plant life that sustains aquatic ecosystems. Sulfuric acid, and metals such as copper, cadmium, lead, zinc, and aluminum damage fish organs and interfere with such functions as growth, reproduction, and respiration. Sediments washed from mines destroy fish habitat or cause direct suffocation of fish and other aquatic animals.

Some of the most dramatic incidents of damage from mining pollution have been massive fish kills, which often occur after a major spill or sudden storm which adds a huge load of pollutants to a stream. But an even greater impact from mining contamination of water is its longer-term chronic degradation of the aquatic environment. For example, at the lowest level of the food chain, organisms such as algae supply nutrition for stream insects. These lower-level organisms are often very susceptible to the toxic effects of acid and metals. Insects are typically prey for larger stream animals such as fish, frogs, and crustacea. These predators decline in number as they take in metals through their food or as their food sources diminish. The result of such contamination is an ailing ecosystem depleted of most aquatic life. In many acid mine drainage-impacted streams, there may be almost no life for several miles downstream of a mine, except for the most acid-resistant species.

PHOTO: ANN MAEST

This brown pelican was found dead on copper encrusted sands where Peru's biggest copper producer, Southern Peru Copper Corporation(SPCC), is responsible for pollution. SPCC is majority owned by U.S. interests including Newmont Mining Corp., Phelps Dodge Corp., the Pritzker family of Chicago and ASARCO Inc. (see Chapter 9, Ilo, Peru).

The casualties of mining's pollution also include land animals dependent on the aquatic environment for food and habitat. Evidence suggests that population levels of species such as mink and otter are reduced near mine-impacted streams in comparison with populations of these species near non-impacted streams.

Cyanide from gold mining operations has also wreaked havoc on aquatic life. One cyanide spill, for example, killed 11,000 fish in South Carolina in October 1990. Birds that mistake cyanide ponds at gold mines for attractive feeding and nesting areas have also been frequent victims of cyanide. Throughout the nineteen-eighties, this was a serious problem in Nevada, where the desert landscape made these cyanide pools a lure to bird life and miners were slow to put covers or netting over their impoundments.

- During one three-month period in 1986, a 20-acre tailings pond at the Paradise Peak Gold Mine, 100 miles southeast of Reno, Nevada, became a lethal oasis for 735 birds.[73] As of 1989, more than 1,600 bird deaths were reported at this mine.[74]
- In November 1995, 340 geese migrating south from Alaska and Canada were found dead and floating in the abandoned Berkeley Pit copper mine in Butte, Montana.[75] Since the now defunct Anaconda Copper Mining Company stopped mining the pit in 1982, and stopped pumping it dry, the massive open pit had filled with groundwater and mine waste contaminants (iron, aluminum, sulfates, copper, arsenic, cadmium).[76] The pit water is now considered a toxic stew. Every day, five million gallons of highly acidic, metal-laden water pour into the pit from the surface and through a network of old underground mine tunnels.[77]
- In 1995, 40,000 gallons of cyanide solution from the New Gold Mine were released into Golconda Creek near Jefferson City, Montana. Surveys conducted after the spill found the creek was devoid of fish, and its population of aquatic insects was greatly reduced.[78]

Impacts on People

Mining is a dangerous and unhealthy business. Mining generates and uses vast amounts of some of the most toxic substances known. Chapter 7 provides disturbing data regarding the effects of this activity on human health. Often, impacts on human populations from mining's water contamination are mixed with those from airborne heavy-metals from smelters, and other pollution sources, so the water-specific effects are hard to isolate. Also, public attention to such impacts in the past has generally been slight because many mining facilities, particularly in the West, are located in remote areas, limiting direct human contact.

However, this provides no comfort for the future. Mining is expanding into many new locations, and the growing population of the West means many once-remote sites now abut significant human communities. The issue of human exposure to mining pollution demands more study and concern.

Some statistical studies of human health in mining districts have been carried out which strongly suggest that there are links between exposure to mining contaminants and human mortality and disease. For example, the death rate from serious disease has been unusually high in the Clark Fork basin near Butte, Montana, an area of intensive mining and smelting for over a hundred years.[79,80] (The Clark Fork basin contains the most extensive area of Superfund sites in the United States.) National cancer statistics have also shown elevated death rates from cancer — particularly lung, bronchial, and trachea cancer — in areas of the Clark Fork basin where mining has occurred. Cancer mortality rates in these areas have been much higher than in other areas in Montana and neighboring states where mining activity has not occurred[81,82] (see Chapter 7, Effects on Human Health).

While these statistics do not separate public health impacts associated with airborne contamination, primarily caused by smelting, from those impacts associated with waterborne contamination, the known higher incidence of mortality and cancers in this area of intensive mining clearly demonstrates cause for alarm, and underscores the need for further research on the health impacts of mining activities on humans.

It is well known that many typical constituents of mining waste can pose serious risks to human health. For example, some mining-released metals that are highly toxic to fish and wildlife are also toxic to humans, including mercury, arsenic, cadmium, chromium, and lead. Long-term exposure to arsenic is associated with skin cancer and other organ tumors, while cadmium exposure can cause kidney disease. Lead can retard normal growth and development in children, and mercury, depending on its form, may cause nervous system damage and gastrointestinal problems. Iron, zinc, and copper, often toxic to fish and insects at very low levels, are toxic to humans at higher levels. Sulfates (nonmetals mobilized during acid mine drainage), when consumed by humans, particularly children, in high doses can cause diarrhea and other gastrointestinal problems[83] (see Appendix).

The public has been exposed to dangerous levels of heavy metals and other contaminants in water near active and historic mines throughout the United

States. In some cases, the levels of metals in drinking water have exceeded the health-based *Maximum Contaminant Levels (MCLs)* established under the federal *Safe Drinking Water Act*[84] (see Table in this chapter). This has usually occurred where people draw their drinking water from wells supplied by an aquifer contaminated by mining operations.

- The Bingham Canyon Mine near Salt Lake City, Utah, the world's largest open-pit copper mine, has generated a massive contaminated plume (a concentrated area of pollution) of groundwater that covers more than 70 square miles downgradient from the mine, threatening the water supply of a Salt Lake City suburb of 70,000 residents. The plume is contaminated with sulfates (from acid mine drainage) and heavy metals. After years of foot dragging and unsuccessful court battles aimed at evading responsibility, the Kennecott Copper Company (current mine operator) has begun taking pollution control efforts to stop further pollution from entering the groundwater. However, to date the company has no plan for cleaning up the existing groundwater pollution, other than to provide alternative water sources to qualified residents[85] (see Chapter 5, Photo).

Cyanide used in gold mining is also of serious human health concern. Small quantities of cyanide cause rapid death. There are some mitigating factors: cyanide breaks down naturally when exposed to sunlight or acidic conditions. Cyanide is not known to bioaccumulate in animal tissues. However, experiments have found that cyanide can persist for at least a century in groundwater and in mine tailings or abandoned leach heaps, particularly where alkaline conditions are maintained.[86] And given its acute toxicity, cyanide remains a significant threat to human health.

While human exposure to sub-lethal concentrations of cyanide have been assumed to be harmless, research has not conclusively determined the effects of long-term, low-level exposure to cyanide on humans. Correlations have been observed between chronic low-level cyanide uptake and specific diseases in humans, and animals exposed to cyanide have shown progressive damage to the nervous system and animal tissues.[87, 88] Clearly more studies are warranted.

The mining industry has not exhibited a level of care commensurate with the risks cyanide poses to water resources. Numerous leaks of cyanide solution from ore heaps have occurred when the liners underneath have torn under the weight of materials or when solution ponds have overflowed after rains. At present, because of insufficient monitoring, we simply do not know how many plumes of cyanide-contaminated groundwater are creeping steadily from mining sites toward water supplies used by communities.

- Five community drinking water wells were polluted with cyanide in 1983 from the Golden Sunlight Gold Mine, near Whitehall, Montana.[89] The source of the pollution was an unlined tailings pond that leaked 160,000 gallons of cyanide-laced mine process water into the ground.[90] In 1986, 2,000 more gallons of cyanide leaked off the site and were blamed for killing five cattle[91] (see Chapter 7, Interview).

National Water Quality Standards for Selected Mining Contaminants

Contaminants	Aquatic Life Standards Criterion Maximum Concentration (μg/L)†	Primary Drinking Standards MCL (μg/L)
Antimony**	-	6
Arsenic	360	50
Beryllium**	-	4
Cadmium	3.7	5
Chromium (III)	550	100
Chromium (VI)	15	100
Copper	17	al 1300
Iron*	-	-
Lead	65	al 15
Mercury	2.1	2
Nickel	1400	‡
Selenium	20	50
Silver*	3.4	-
Thallium**	-	2
Zinc*	110	-
Cyanide	22	200

† = The highest concentration of a pollutant to which aquatic life can be exposed for a short period of time (1 hour average) without deleterious effects.

al = action level (Water systems must take action if more than 10% of their tap water samples are above that level.)

* = Iron, Zinc, and Silver are contaminants listed under the National Secondary Drinking Water Standards. The suggested level under these standards is 300 μg/L, 5000 μg/L, 100 μg/L respectively. For Aquatic Life Standards, Iron is considered a non-priority pollutant. The criterion maximum concentration is 82 μg/L for which the hardness dependent criteria is 0.1 μg/L.

‡ = The MCL for Nickel was remanded as of February 23, 1995 until an appropriate methodology for deriving the maximum contaminant level goal is determined (Federal Register, 60 (125), Rule and Regulations, 29 June 1995).

** = Antimony, Beryllium, and Thallium are not considered priority pollutants as published in 40 CFR 131.36. The proposed criterion for Antimony is 88 μg/L. For Beryllium and Thallium, there is insufficient data to develop criteria, but the lowest observed effect levels (LOEL) are 130 μg/L and 1400 μg/L respectively.

Sources:

For Aquatic Life Freshwater Criterion Maximum Concentration Standards:
United States Environmental Protection Agency, 40 CFR 131.36 (7-1-95 Edition)

For National Primary Drinking Water Standards:
United States Environmental Protection Agency, Safe Drinking Water Hotline, National Primary Drinking Water Standards (February, 1994).

- In late 1994, cyanide was discovered in a residential drinking water well below the Pony Mill (gold ore processing facility) in Pony, Montana. The cyanide was leaking from the mill's wastewater ponds, located directly upgrade from the town of Pony.[92] In 1989, residents had warned regulators of such potential problems prior to the mill's opening, but the State of Montana ignored these warnings (see Chapter 7, Photo).

Besides affecting drinking water supplies, mining pollution may make streams and rivers unfishable, taking a toll not only on recreation but also on local recreation-based economies. Pollution and use of water may also interfere with agriculture and other industries. Thus, the economic consequences for humans of mining's damage to water are important alongside the industry's damage to health.

- Aquatic life in Missouri's Big River, formerly one of the state's most popular fishing streams, has been severely damaged by a tailings dam failure at a closed lead mine in the early 1980s. Heavy rains in 1993 flushed more tailings waste into the river.[93] Because of lead contamination of fish, the state has issued a health advisory against eating fish caught along a 70-mile length of the Big River.[94]

- In California, state advisories currently warn against eating any fish from six different reservoirs and streams because of elevated mercury levels in fish. Other advisories recommend limiting fish consumption from five other bodies of water.[95] Past mercury mining in the state has contributed to this contamination.[96]

Interview:

Andrea Fine

Georgia Heartbreak

by Andrea Fine

Andrea Fine is currently a freelance journalist working in Philadelphia, Pennsylvania. Ms. Fine is a former newspaper reporter and served as a correspondent for **People Magazine.**

In 1968, C.E. Minerals (now called Combustion Engineering, Inc.) began to mine Graves Mountain near Lincolnton, Georgia, for kyanite. Kyanite, a blue aluminum silicate mineral, is used to make refractories or heat-resistant materials, such as for NASA's Space Shuttle program.

Kyanite mining at the Graves Mountain mine was similar to gold mining. Huge sections of the earth were dug up, the ore was processed to separate out the kyanite, and huge tailings piles of mining wastes were left behind. Over time, pyrite and sulfur waste materials, exposed through the mining process, were discarded on the tailings piles. The waste oxidized, and rain and mountain spring water converted this material into an acid generating mess that washed into nearby streams. In about 1985, the mine closed. The company made little attempt at remediation or reclamation.

PHOTO: HOWIE OAKES

Raleigh Long, Sr. stands beside the contaminated stream that runs through his property.

Raleigh Long, Sr., a 78-year-old retired builder, owns a 200-acre cattle farm about three miles south of the mine site. One of the contaminated streams flows through his property. By the mid-1970s, that stream had died from an over-saturation of acid and heavy metals, including manganese, aluminum, and iron. Long's cows refused to drink from the stream; aquatic life died out. He feels his property was significantly devalued.

Long wasn't the only landowner affected by the acid mine drainage. Ultimately 15 other community residents filed a class-action lawsuit against the company, in the case of *Johansen, et al. vs Combustion Engineering, Inc.* A jury ruled in favor of the landowners and awarded them punitive damages of $45 million — which the judge later reduced to $15 million. Combustion Engineering appealed the decision to the Federal 11th Circuit Court of Appeals in Atlanta. In September 1995, the Circuit Court upheld the lower court's decision, but later that year, the company petitioned the U.S. Supreme Court to hear the case. In May 1996, the Supreme Court, without rendering a decision, remanded (returned) the case to the Circuit Court for further legal analysis. As of 1997, the case is still in limbo, and the plaintiffs have not yet recovered any punitive damages.

But for Long, the issue is the land and water, not the money. In his own words, he tries to make sense of the tragedy:

Most of the refuse they cast off from the mine came right through my place. It killed all the ecology in the stream — all the fish, all the snakes and the aquatic life. It's still dead at the present time; there's no way they can reclaim it or cover it up. They left an ongoing sore, a cancer, so to speak, that's ongoing right now.

We haven't seen any money from the court case. We haven't seen anything, and I don't guess we ever will. What I'm most interested in is the creek, but I don't see any solution. This mine has exposed the whole side of the mountain. You used to see a lot of aquatic life down there, and you used to fish in it and you used to swim in it. It's not even something you can wash your feet in anymore. I have tributaries that run into the stream where the water's still fresh. If it wasn't for that, I wouldn't have any water at all and neither would my family downstream from me.

I'm an old fashioned man — I like clean water.

Past, Present, and Future

Mining which has been carried out in the past has caused massive water contamination around the United States, and the world. The forms and pathways of that contamination have been outlined in this Chapter.

Protecting water resources from mining damage is not just a problem of the past, however. While some of the worst mining practices that damaged water, such as open dumping of tailings into rivers, are now outlawed in the United States, those practices endure in other parts of the world (see Chapter 9).

And even in the United States, important pathways of water damage are not regulated: there is no national groundwater protection program, and no general standards to ensure that hardrock mining operations are managed to prevent groundwater contamination.

This means that water contamination from mining is not simply a problem of the past. It continues today. And as the world's economic development proceeds, our global consumption of metals accelerates. Mining is accelerating to match that demand, and mining's impact on water resources is accelerating too. Unfortunately, despite the growth in environmental concern of the last thirty years, the rise in mining impacts is outpacing our controls on those impacts.

We can solve these problems. Though mining imperils water supplies, properly managed mines can operate without spreading contamination. Chapter 4 describes some of the techniques that have been developed to meet this need.

Mining's effects on the environment last a long time. Mining will endanger public health, the natural environment, and the quality of life in an increasing number of communities if steps are not taken to upgrade mine design, environmental monitoring, waste management, and reclamation practices. If mines are allowed to continue practices that imperil water resources, the results will be devastating. The time for action is now.

♦ ♦ ♦

Notes

1. U.S. ENVIRONMENTAL PROTECTION AGENCY, MINING WASTE NPL SITE SUMMARY REPORT, CALIFORNIA GULCH, LEADVILLE, COLORADO, p.2 (21 June 1991).

2. David Fanning, *Bleeding From a Million Wounds*, HIGH COUNTRY NEWS, 20 November 1989, p.11.

3. *Id.*

4. MINING WASTE NPL SITE SUMMARY REPORT, *supra* note 1, p.9.

5. *Id.*, p.10.

6. *Id.*

7. U.S. GEOLOGICAL SURVEY, NATIONAL OVERVIEW OF ABANDONED MINE LAND SITES UTILIZING THE MINERALS AVAILABILITY SYSTEM (MAS) AND GEOGRAPHIC INFORMATION SYSTEM (GIS) TECHNOLOGY, pp.15-16 (1996).

8. *U.S. Takes Idaho Mining Companies to Court*, CASCADIA, May 1996, p.6.

9. Susan Drumheller, *Floods Bring Million Pounds of Lead to Lake*, SPOKESMAN-REVIEW, 13 June 1996, p.1.

10. Mark Matthews, *Idaho's Crown Jewel Set in Deadly Metals*, THE WASHINGTON POST, 28 December 1996, p.A3.

11. *U.S. Takes Idaho Mining Companies to Court*, *supra* note 8, p.6.

12. DENNIS SLIFER, NEW MEXICO ENVIRONMENT DEPARTMENT, EXECUTIVE SUMMARY, RED RIVER GROUNDWATER INVESTIGATION, pp.1-3 (March 1996).

13. Concerned Citizens of Questa, New Mexico, Testimony to the House Interior Subcommittee on Mining and Natural Resources (18 April 1989).

14. Bill Harlan, *Leach Mining Killing Birds*, RAPID CITY JOURNAL, 28 December 1992, p.1.

15. George Ledbetter, *Wharf Fined $150,000 for Release*, THE LAWRENCE COUNTY CENTENNIAL, 21 October 1995.

16. Telephone communication with Hugh Ricci, Nevada State Engineer's Office (7 February 1997).

17. Iron sulfides such as pyrite and marcasite frequently occur in coal bodies, where they produce acid mine drainage in the Eastern United States.

18. PROCEEDINGS, FIRST MIDWESTERN REGION CONFERENCE (held at Southern Illinois University at Carbondale), p.1–1 (June, 1990).

19. Charles N. Alpers & Darrell Kirk Nordstrom, *Geochemical Evolution of Extremely Acid Mine Waters At Iron Mountain, California: Are There any Lower Limits To pH? in* PROCEEDINGS, SECOND INTERNATIONAL CONFERENCE ON THE ABATEMENT OF ACIDIC DRAINAGE, pp.321-342 (1991).

NOTES

20. THE UNIVERSITY OF CALIFORNIA AT BERKELEY, MINING WASTE STUDY, FINAL REPORT, p.262 (1988).

21. *Id.*, p.15.

22. *Id.*

23. HYDROMETRICS INC., FISHER CREEK POLLUTION LOADING ANALYSIS (Prepared for Douglas Parker, Crown Butte Mines Corp., Missoula, MT), (August 1993).

24. TIMES LTD. INC. ET AL., INTERGRATED WATER MANAGEMENT AND DISCHARGE STRATEGY FOR THE NEW WORLD MINE (Prepared for Crown Butte Mines Inc., Missoula, MT), (October, 1994).

25. U.S. ENVIRONMENTAL PROTECTION AGENCY, ACID MINE DRAINAGE PREDICTION (DRAFT), pp.18-19 (1994).

26. Telephone communication with Steve Hoffman, Office of Solid Waste, Special Waste Branch, U.S. Environmental Protection Agency (22 January 1997).

27. Johnnie N. Moore & Samuel N. Luoma, *Hazardous Wastes From Large-scale Metal Extraction,* ENVIRONMENTAL SCIENCE AND TECHNOLOGY, p.1279 (1990).

28. John Holt, *Killing Clark Fork Browns*, FLY FISHERMAN, July 1990, p.17.

29. U.S. ENVIRONMENTAL PROTECTION AGENCY, MINING SITES ON THE NPL (DRAFT), p.13 (June, 1996).

30. Telephone communication with Shawn Hogan, Remedial Project Manager, Carson River Superfund Site (24 July 1995).

31. *Id.*

32. Division of Environmental Protection, State of Nevada Department of Conservation and Natural Resources.

33. Mark Obmascik, *Montana Mine Had 9 Cyanide Leaks,* THE DENVER POST, 2 November 1993, p.4A.

34. DAMES & MOORE, PTFI ENVIRONMENTAL AUDIT REPORT p.4 (25 March 1996).

35. *Id.*, p.36.

36. AMERICAN MINES HANDBOOK 1995, p.178 (Southam Magazine & Information Group 1994).

37. U.S. DEPARTMENT OF THE INTERIOR & STATE OF MONTANA DEPARTMENT OF ENVIRONMENTAL QUALITY, DRAFT ENVIRONMENTAL IMPACT STATEMENT FOR ZORTMAN AND LANDUSKY MINES, pp.2-32, 2-59 (August, 1995).

38. Mark Obmascik, *Montana Mine Had 9 Cyanide Leaks,* THE DENVER POST, 2 November 1993, p.4A.

NOTES

39. Memo from M.K. Botz, Hydrometrics, to File (undated). (Memo references cyanide contamination in Alder Gulch and Kalal water supply.)

40. Stipulation and Agreement between Pegasus Gold Inc. and Montana Department of Health and Environmental Sciences (1 April 1983).

41. Letter from John Fitzpatrick, Pegasus Gold Corp., to Sandi Olsen, Montana Department of State Lands (14 June 1991). Letter notes that mine manager intends to net ponds to prevent recurrence of accident.

42. Obmascik, *supra* note 38 p.4A.

43. Children tested by William LiPero, M.D., January, 1986. Lead concentrations were analyzed by MetPath, Teterboro, N.J. (October, 1992-March, 1992).

44. SCOTT HAIGHT & JOE FRAZIER, U.S. BUREAU OF LAND MANAGEMENT AND SCOTT SPANO, MONTANA DEPARTMENT OF STATE LANDS, EMERGENCY TREATMENT AND LAND APPLICATION OF EXCESS CYANIDE SOLUTION OF THE ZORTMAN MINE, PHILLIPS COUNTY, MONTANA (undated).

45. Letter from Scott Haight et al., U.S. Bureau of Land Management, to Mike Irish, Montana Department of State Lands, p.1 (5 October 1987).

46. Obmascik, *supra* note 38 p.4A.

47. Jill Sundby, *EPA Cites Zortman-Landusky Mine,* THE BILLINGS GAZETTE, 13 August 1993, p.C1.

48. Jill Sundby, *State Sues Zortman Mine,* THE BILLINGS GAZETTE, 27 August 1993, p.5C.

49. Nick Ehli, EPA *Sues Zortman Mining,* THE BILLINGS GAZETTE, 7 June 1995, p.1A.

50. Claire Johnson, *Paying the Price,* THE BILLINGS GAZETTE, 23 July 1996, p.1A.

51. U.S. BUREAU OF MINES, WATER USE IN THE DOMESTIC NONFUEL MINERALS INDUSTRY, Information Circular 9196, p.7 (1988).

52. *Id.*

53. ARIZONA DEPARTMENT OF WATER RESOURCES, TUSCON ACTIVE MANAGEMENT AREA, SECOND MANAGEMENT PLAN 1990-2000, p.184 (1988).

54. Telephone communication with Daniel Edelstein, copper commodities expert, U.S. Bureau of Mines (7 February 1996).

55. In 1994, the city of Austin, Texas used 37.5 billion gallons of water or 102.7 million gallons per day. TEXAS WATER DEVELOPMENT BOARD, HISTORICAL SUMMARY OF CITY WATER USE (1995).

56. Id., p.17.

57. Id., p.35.

NOTES

58. Lane White and Steve Kral, American Barrick, MINING ENGINEERING, p.1231 (November, 1994).

59. TEXAS WATER DEVELOPMENT BOARD, HISTORICAL SUMMARY OF CITY WATER USE (1995). At 70,000 gallons of water pumped per minute, the Goldstrike Mine uses 103.2 million gallons of water per day.

60. E.J. CROMPTON, U.S. GEOLOGICAL SURVEY, POTENTIAL HYDROLOGIC EFFECTS OF MINING IN THE HUMBOLDT RIVER BASIN, NORTHERN NEVADA, WRIR 94-4233 (1995).

61. In 1995, 161,000 acre/feet of water were used for domestic purposes in greater Seattle, Washington (population 1.1 million). Telephone communication with Ron Lane, Water Resources, United States Geological Survey (27 January 1997).

62. CROMPTON, *supra* note 60.

63. GLENN C. MILLER, MINE DEWATERING DEFICITS IN THE HUMBOLDT RIVER BASIN: ESTIMATES OF VOLUME AND ASSOCIATED IMPACTS (DRAFT), p.3 (1996).

64. Conversation with Tom Leshendock, Deputy State Director, Mineral Resources, U.S. Bureau of Land Management (11 May 1995).

65. Telephone communication with Hugh Ricci, Nevada State Engineer's Office (24 January 1997).

66. *Id.*

67. Dorothy Y. Kosich, Special Interests Take Aim at Pipeline, MINING WORLD NEWS, April/May 1996, p.2.

68. TOM MYERS, CUMULATIVE HYDROLOGIC EFFECTS OF OPEN PIT GOLD MINING IN THE HUMBOLDT RIVER DRAINAGE, A REPORT TO THE SIERRA CLUB, p.17 (10 March 1994).

69. *Id.*, p.17.

70. CROMPTON, *supra* note 60.

71. ARIZONA DEPARTMENT OF WATER RESOURCES, 1984-1994 ANNUAL WATER WITHDRAWAL AND USE SUMMARY (27 November 1995).

72. ARIZONA DEPARTMENT OF WATER RESOURCES, INDUSTRIAL WATER USE SUMMARY (1994 & 1996).

73. Tom Harris, *U.S. Launches Criminal Probe into Desert Pond Bird Deaths,* THE SACRAMENTO BEE, 27 June 1986, p.B8.

74. Doug McMillan, *Miners Try to Evict Birds From Cyanide Ponds,* RENO GAZETTE-JOURNAL, 2 April 1989, p.3D.

75. Erin P. Billings, *Berkeley Pit Water Kills Healthy Birds in ARCO's Tests,* HELENA INDEPENDENT RECORD, 26 April 1996.

76. Andy Davis & Daniel Ashenberg, *The Aqueous Geochemistry of the Berkeley Pit, Butte, Montana, U.S.A., in* APPLIED GEOCHEMISTRY, Vol. 4, pp.23-36 (1989).

NOTES

77. CANONIE ENVIRONMENTAL SERVICES CORP., DRAFT REMEDIAL INVESTIGATION REPORT, BUTTE MINE FLOODING OPERABLE UNIT (Prepared for ARCO), p.134 (January, 1994).

78. *A Year of Mining Crimes,* CLARK FORK-PEND OREILLE COALITION CURRENTS, November/December 1995, p.8.

79. H.I. SAUER & L.E. REED, TRACE SUBSTANCE IN ENVIRONMENTAL HEALTH, p.62 (1978), cited in Moore & Luoma, *supra* note 27.

80. M. FEINLEIB, ET AL., MORTALITY FROM CARDIOVASCULAR AND NON-CARDIOVASCULAR DISEASES FOR U.S. CITIES (NIH Publication No. 79-1453, 1979), *cited in* Moore and Luoma, *supra* note 27.

81. T.J. MASON, ET AL., THE 1975 ATLAS FOR CANCER MORTALITY FOR U.S. COUNTIES, 1950-1969 (NIH Publication No. 75-780, 1976) U.S. NATIONAL CANCER INSTITUTE, DEPARTMENT OF HEALTH, EDUCATION, AND WELFARE, *cited in* Moore and Luoma, *supra* note 27.

82. W.B. RIGGAN ET AL., U.S. CANCER MORTALITY RATES AND TRENDS, 1950-1979, U.S. EPA PUBLICATION #EPA-600/1-83-0156 (1983), *cited in* Moore and Luoma, *supra* note 79.

83. U.S. ENVIRONMENTAL PROTECTION AGENCY, FACT SHEET: NATIONAL SECONDARY DRINKING WATER STANDARDS (undated).

84. 33 U.S.C.A. §§ 1251 to 1387.

85. Telephone communication with Eva Hoffman, Remedial Project Manager for Kennecott Superfund site (27 January 1997).

86. P.R. Engelhardt, *Long-Term Degradation of Cyanide in an Inactive Leach Heap, in* CYANIDE AND THE ENVIRONMENT, p.539 (1985).

87. OAK RIDGE NATIONAL LABORATORY, REVIEWS OF THE ENVIRONMENTAL EFFECTS OF POLLUTANTS: V. CYANIDE, pp.139-145 (1978).

88. LEWIS R. GOLDFRANK ET AL., GOLDFRANK'S TOXICOLOGICAL EMERGENCIES, 3rd Ed., p.587 (1986).

89. HARDROCK BUREAU, MONTANA DEPARTMENT OF STATELANDS, INVESTIGATION OF GOLDEN SUNLIGHT MINE'S TAILINGS POND LEAK AND ALLEGED IMPACT TO DOWNGRADIENT DOMESTIC WATER SUPPLIES, p.4 (15 May 1987).

90. UNITED STATES COMMITTEE ON LARGE DAMS, TAILINGS DAM INCIDENTS, p.46 (November, 1994).

91. Memorandum from Gary Lynch, Montana Department of State Lands to Terry Grotbo, Montana Department of State Lands (9 May 1986).

92. *Cyanide Leak out of Pond Irks Residents,* 8 December 1994, BILLINGS GAZETTE.

93. Doug Hawes-Davis, *Mining the Ozark Highland,* FOCUS, p.9 (1990).

NOTES

94. Ted Williams, *The Price of Lead*, AUDUBON, January-February 1995, p.31.

95. CALIFORNIA DEPARTMENT OF FISH AND GAME, CALIFORNIA SPORT FISHING REGULATIONS, pp.11-12 (1994).

96. Telephone communication with Rick Humphries, California Water Quality Control Board (7 February 1997).

♦ ♦ ♦

We Know How To Stop Mining Pollution

4

Polluted water is not an inevitable "price we must pay" to use metals in our lives. We have the scientific knowledge and the practical technology, to prevent water pollution from mining.

Metals and mineral products play a vital role in modern civilization. No one proposes that we stop using these resources. But, just because we need metals does not mean we must accept polluted water. We can have one without the other.

Just as our scientific understanding of the toxic damage that heavy metals like mercury, lead, and copper have on the health of human beings and natural systems has advanced in recent decades, so has our knowledge of mining methods to prevent pollution. Today, the mining industry has the means to design and operate clean mines to prevent hazards such as acid mine drainage, metals release, and stream sedimentation. The industry can produce minerals profitably while utilizing these methods. Environmentally-sensitive mining is ultimately in the industry's financial interest because it can spare mine operators unnecessary cleanup, remediation expense, and legal liability. But, while a limited number of mines have adopted these techniques, most mines do not employ them consistently, if at all.

Improving mining's environmental performance will require more than adopting a set of technical and engineering fixes, however. Mining companies also must foster a commitment to environmental excellence in their entire workforce — from senior management to on-the-ground operators. This commitment by management has to be expressed in company policies that determine personnel evaluation policies and promotions, to have a real impact on the mine site.

Yet, history has shown that in the end we cannot entrust the protection of the public's water resources to the private interests of mining companies alone. The federal government and the states must adopt effective laws, mines must be monitored on a regular basis, and companies must post financial guarantees that will cover the cost of reclamation and long-term treatment if water pollution occurs. Polluting mines must be penalized, so their management cannot make a profit from environmental neglect. For the thousands of abandoned mines which are currently polluting water resources, a nationwide hardrock abandoned mine reclamation program must be established.

Techniques to Prevent and Treat Water Contamination from Mining

Mines operators can protect water resources by adopting a number of practices to prevent, control, and clean up mining-caused pollution.

The environmentally-responsible mine operator should do everything possible to prevent water contamination at the outset, through proper mine planning and management of wastes. A key prevention goal is to isolate sulfidic wastes from water and oxygen, because these two substances trigger the generation of acid mine drainage. Halting the initial oxidation of sulfides is very important because, if it is allowed to progress, acid generation becomes greatly accelerated by bacterial action which makes the process of acid generation very difficult — and expensive — to control (see Chapter 3, Acid Mine Drainage). Separating mine wastes from water is also essential because water is the chief mechanism which transports acid mine drainage, sediments, and processing chemicals through the environment. These practices also must be incorporated into the process of restoring mine-damaged land to a productive use (reclamation).

Treatment of contaminated water, while often necessary, should be the miner's last resort. Treatment often must be perpetual and is considerably more expensive than measures to prevent mining pollution. Perpetual treatment requires perpetual funding.

As the sections below describe, techniques to prevent and treat pollution from mining are most successful if they are part of an integrated pollution control plan that takes effect before mining begins.

Environmental Planning

Proper planning before mining begins is critical in locating and designing a mine to prevent the discharge and migration of acid, metals, sediments, cyanide and other chemicals. If location and design are given short shrift and pollution problems develop, the required cleanup actions are almost always

PHOTO: STEVE SHIRLEY

The abandoned Nellie Grant Mine, southwest of Helena, Montana, underscores the disastrous environmental and financial impacts of poorly planned and mismanaged mines.

more expensive and more complicated than the work to prevent that pollution would have been. Poor location and design work can produce pollution time bombs, such as acid mine drainage which develops at a site years after the mine ceases operation. Once they develop, these problems can be impossible to stop.

Preventing pollution requires a multi-pronged approach. To begin, a mine operator must compile detailed information about the site's terrain, climate, surface and groundwater hydrology, and geochemistry (see Chapter 5, Better Baseline Studies). With this information, the mine can be designed to minimize risk. Areas where rock contains high levels of acid-forming sulfides and relatively little buffering material present a high risk of acid mine drainage. Areas with heavy water drainage may also have a high level of risk. If the combination of risks at a particular site is so high that measures to protect water may not work, a mine should not be opened.

The prudent operator will design and locate mine facilities such as waste rock dumps, tailings impoundments, and leach heaps to minimize their contact with water over the course of a mine's life and after mine closure. Avoiding contact between water and wastes in drainage areas will help prevent acid mine drainage, metal transport, stream sedimentation, and pollution from processing chemicals. As a rule, waste facilities should be located in "high and dry" areas outside of the paths of flowing waters, and removed from springs.[1]

A mine operator also must incorporate knowledge of local climate and weather conditions, such as precipitation and evaporation, into mine planning. The operator must design facilities not only to accommodate the water that is required in normal mineral processing but the water that may be introduced into the mine operation through rain, snowmelt, and runoff. Waste impoundments or leach ponds, for instance, must be built properly to ensure stability, and they must be large enough to contain the excess water that accumulates after extreme weather events like rainstorms and snowmelts. Otherwise, drainage from large storms and seasonal snowmelt may breach impoundment dams or cause overflows that carry huge quantities of mining contaminants to surface and groundwater.

Knowing the chemical composition of the ore and waste rock that will be encountered in mining is an essential part of mine planning. Before mining, the operator should implement a thorough system of rock sampling and testing to determine the acid generation potential of the range of rock types that will be encountered in mining. As discussed in Chapter 3, simple reliance on acid-base accounting (comparing bulk amounts of acid-generating and acid-neutralizing constituents in mining wastes) often has led to errors in predicting acid mine drainage. Instead, the operator should use kinetic testing methods to assess the potential for acid generation in response to environmental factors — such as air temperature, moisture, and bacterial action — that are likely to be present at the mine site. Accurate information on the likelihood of a particular waste type to generate acid also can allow the mine operator to handle riskier waste types separately by isolating them from the elements.

Many of the worst cases of mining pollution can be traced to poor or simplistic planning and assessment. Planning may not result in a perfect mining operation, but it will provide a better one. Carried out properly, planning will eliminate many potential pollution problems.

Managing Wastes to Prevent Water Contamination

Handling mine waste conscientiously is essential to prevent water contamination from mining, thereby making expensive pollution treatment methods unnecessary. The most effective way of reducing the risk of water contamination from mining is to isolate mining wastes from the environment. Another promising, though less reliable technique involves combining or blending potentially acid-generating wastes with materials that can buffer or neutralize the acidity that is generated in wastes.

As the following sections describe, many effective waste handling techniques involve simple, straightforward technology. Operators can perform many of these techniques with commonly available materials; the materials are often available locally, or even at the mine sites themselves.

Isolating and Containing Wastes

Isolating and containing mining wastes from the elements confers two benefits. First, it can inhibit the initial formation of acid mine drainage by preventing the contact of sulfide material with oxygen and water. Second, waste containment can prevent the migration of pollutants such as acid, metals, and sediments into the environment.

The appropriate degree of isolation and containment will depend in part on the stage of the mining operation. While an operator is actively depositing waste into a tailings impoundment or onto a waste rock pile, the extent of possible isolation and containment is more limited than when an area of waste disposal is no longer receiving new waste. During periods of active waste disposal and ore heaping, the mine operator must ensure that a waste dump or heap leach pile be structurally intact. The area must be properly lined at its base, so contaminated water cannot leak into the ground and contaminate aquifers. Any natural waterflows in the area must be permanently diverted away from the mined material.

When waste disposal areas are filled up and retired, they can be more completely isolated by methods such as capping the waste material. Submerging wastes into water (as examined later) is another isolation technique, but it is experimental and carries substantial environmental risk.

There are a number of basic methods in current practice that can be utilized to contain mining wastes, and to isolate them from water and oxygen. The methods outlined below have records of cost-effectiveness and success:

Sound Tailings Impoundment Design — When the marketable metals or other products are extracted from bulk ores, large amounts of ground-up waste rock, or tailings, are left behind as waste. These mining tailings were once simply dumped into the nearest river. Today they are piped as a stream of thick, muddy water into specially-built ponds, called tailings impoundments. Tailings impoundments, which typically store millions of tons of unstable, water-saturated wastes, can pose a serious risk of contamination, either through leakage into groundwater or through massive structural failure of the dams. Ensuring that tailings impoundments are structurally intact, and preventing contact with water drainages, are important steps which must be followed to reduce the risk of water contamination both during and after mining.

The practice of storing mine wastes in tailings ponds and impoundments has been utilized for centuries. Unfortunately, history has shown us that tailings impoundments often fail to contain their toxic wastes due to poor design, improper construction, negligent management, or inadequate inspec-

PHOTO: PHILIP M. HOCKER/ MINERAL POLICY CENTER

Tailings impoundments and leach heaps must be properly lined and maintained to prevent contamination of groundwater. In this photo, a lined cyanide solution pond is being filled in the Carlin district, Nevada.

tion and monitoring. Nevertheless, impoundments continue to serve as the primary method for containing mine tailings. When properly constructed and maintained, they can prevent serious environmental contamination.

Attention to several key issues will help ensure a sound impoundment system. First, the impoundment must be located in an area that has a stable foundation, and where surface drainage is minimal or can be diverted around the structure permanently. Separating rainfall and streamwater from tailings impoundments can be challenging, because impoundments are often built in valleys or bowl-like depressions. While valleys provide a natural impoundment for tailings, they are also natural drainage areas. The operator therefore must direct all natural water flows away from tailings ponds to prevent contact of surface water with mining waste. It is more expensive to build tailings storage impoundments in upland areas away from natural stream channels, because a longer dam must be built, but it is a better long-term practice.

Impoundments also must have ample holding capacity, not only to hold the maximum load of wastes that will be generated, but also to handle added drainage from large rainstorms and heavy snowmelt. The impoundment design also should include a containment pond below the structure to catch unanticipated spills from the main impoundment. The impoundment must be lined properly to prevent groundwater contamination. The dam or dikes that enclose the impoundment must be designed carefully and constructed using state-of-the-art engineering and materials to ensure its strength and stability, and to prevent leaks.

Finally, mine operators should take steps to reduce or remove toxic substances (cyanide, acid, metals, etc.) in the tailings stream prior to storage in the impoundment. Reducing the toxic nature of the waste reduces the environmental impact that could occur should the impoundment later fail and spill its contents off site. Cyanide concentration levels, in particular, should be reduced inside the mill, before a stream of cyanide-laden tailings is piped into an open pond. This also can minimize the risk of wildlife deaths that could result when animals come into contact with the mine facility.

Waste Rock Storage — Locating new mine waste rock piles and moving existing waste piles to secure places outside of primary water drainage areas is a fundamental requirement for preventing acid mine drainage, toxic metals, and erosion from polluting surface and groundwater. Building diversion ditches around waste rock piles or constructing underdrains consisting of coarse, non-reactive (non-acid generating) rock are also ways to avoid contact between water and potentially acid generating waste.

- Years of toxic drainage from waste rock piles contributed to a massive groundwater pollution plume that extends for 70 square miles at Kennecott Corporation's Bingham Canyon Copper mine near Salt Lake City, Utah (see Chapter 3, Impacts on People). To stop acid and metals from continuing to leach into the groundwater, the company moved more than two million tons of waste rock from locations that serve as natural groundwater recharge areas to less environmentally-risky locations.[2] It would have been far cheaper for the company to have placed the waste piles out of the drainage areas initially, but current remedial action has had a significant impact on reducing pollution flow.

Secure Protective Liners — Requiring a secure, multiple lining system underneath waste rock piles, tailings impoundments, leach heaps, and solution ponds can prevent contaminants from migrating into surface and groundwater. Several states now require that waste impoundments and leach dumps have an underlying synthetic liner or an alternative barrier that is equally effective. Virtually all states require liners under new chemical leach heaps.

Unfortunately, few states detail how these liners are to be constructed and maintained, and poor design, construction error, and improper maintenance can result in liners that leak.[3]

All liners are prone to leaking. It is only a matter of time. Therefore, operators must build redundant, multiple-liner systems to provide a back-up protective shield to prevent leaking contaminants from migrating into the environment.

The preferred liner system consists of two independent synthetic liners (typically of thick plastic) overlying a natural base of low-permeability soil, such as clay. The clay serves as a backup barrier to prevent contaminants that have leaked through the synthetic liners from entering the groundwater. A leak detection system should be installed between the two synthetic layers to signal when leaks occur in the top liner, so that operators can take corrective action. The leak detection system should consist of a free-draining gravel or synthetic honeycomb material, which will allow leaks in the top liner to flow quickly to a sump at the heap's lowest point. Monitors can be lowered into the sump periodically to detect leakage through the top liner. The second synthetic layer, below the leak detection system (and backed up by the underlining foundation of clay), provides a final barrier to prevent contaminants that leak through the first liner from reaching groundwater.[4]

Most synthetic liners used today are made of high-density polyethylene (HDPE), a tough, low-permeability plastic designed to be resistant to stress and degradation by chemicals. HDPE liners are 20 to 100 mils in thickness[5] (a mil is one/one-thousandth of an inch), 8 to 80 times thicker than a typical lawn leaf bag. Liners are manufactured in panels 20 feet wide, and up to one thousand feet long.[6] To line large mine waste facilities, individual panels must be rolled out and seamed together on site. Most panels are joined together using a heat or chemical process.

Operators must take special precautions in the installation and placement of liners. Synthetic liners are prone to leaks and tears, especially along seams. They must be inspected carefully when installed, and monitored closely thereafter. A bed of cushioning material, such as finely ground ore, should be placed over the top synthetic liner to protect it from tearing or puncturing when the ore is loaded on top. Likewise, the clay layer beneath the bottom synthetic liner should be prepared so that it is free of sharp objects that could perforate the liner. Workers also should take special care not to drop sharp tools or operate heavy equipment on the liner before it is protected by the heap.

Manufacturing flaws can damage the integrity of synthetic liners. Small pinholes in liners or weak sections of material can be future points of leakage. Improper handling and storage before installation also can degrade liners. Rough handling can puncture and tear liners, while exposing liners to extremes of heat and cold, rain, or ultraviolet radiation may reduce their effectiveness.[7]

Liners should never be installed in winter. If a liner is installed on a base of frozen ground, the liner's foundation will be unstable and settle when

PHOTO: PHILIP M. HOCKER/ MINERAL POLICY CENTER

Mine operators at the Hayden Hill Mine in California install a lining system. Protective liners at mine sites must be deposited on a smooth foundation and then carefully bonded together.

thawing occurs. When millions of tons of heap-leach ore are stacked on top of the liner, the ore's weight over the thawed ground will stretch and tear the liner. Wintertime construction of a heap leach liner contributed to the massive discharge of cyanide-laced water at the Summitville mine in southern Colorado (see Chapter 1, The Sorry Legacy).

Capping Waste Rock, Tailings, and Heaps — When waste rock dumps, tailings impoundments, and heaps are retired, more complete isolation from the elements is accomplished by covering, or *capping,* the waste. Caps can be constructed by covering the wastes with materials such as clay, soil, or synthetic material (or some combination of these materials). The purpose of the cap is to seal off the waste from air and water, thereby inhibiting the process of acid formation. Capping is also effective as a barrier to prevent erosion of mining wastes and the leaching and migration of acid and contaminants into the environment.[8] Clay is a desirable capping material because of its low permeability and common availability. However, clay must be carefully installed, and it may lose its effectiveness as a water barrier if it is allowed to crack, either through drying or exposure to repeated cycles of freezing and thawing. Therefore, in many situations, the operator should place a protective cover above the clay liner to keep it moist and insulated.

Once capped, a waste facility can be covered with topsoil and revegetated. Plant roots prevent erosion of the waste and add structural stability by anchoring the capping material. They also inhibit infiltration of water into the waste by *evapotranspiration,* a process by which plants emit water into the atmosphere — either by evaporation from plant surfaces, or by water loss through their cell walls and cell openings. To prevent erosion of the protective cap, the operator also must ensure that surface water drainage is permanently diverted away from the closed mine waste repository.

Identification of rock types in the waste can be helpful in designing an effective cap. During mine life, reactive wastes (wastes likely to generate acid) can be segregated, and encapsulated with low-permeability and non-acid generating material available at the mine site. To provide an additional protection against the risk of acid mine drainage, the wastes also can be mixed with alkaline material such as lime, limestone, fly ash, and caustic soda, which can buffer any acidity that is generated in the waste.

Pegasus Gold Inc., at its environmentally-troubled Landusky mine in northern Montana, is now capping potentially acid-generating waste rock to stem a growing acid mine drainage problem at the mine. Under a State of Montana and U.S. Bureau of Land Management approved plan, Pegasus is capping reactive waste rock with a one-foot layer of clay, an overlying three-foot layer of non-reactive waste rock, and a top layer of topsoil. Clay for the cap comes from a nearby clay pit. The layer of non-reactive rock serves as a drainage layer to permit water to drain off the waste rock pile and reduce water pressure on the clay cap. After completing the caps, Pegasus will revegetate the waste rock piles.

Backfilling Open Pits — Mine operators also can isolate environmentally-harmful contaminants by depositing them in open mine pits. This action helps restore the pits to be part of a useful landscape in the future. However, mine waste in pits can become an environmental liability. Special factors must be taken into consideration to ensure that refilling pits is environmentally safe. Installing a layer of impermeable material in the pit to prevent waste contaminants from seeping down into groundwater may be required. In addition, if the bottom of the pit is below the water table, the mine pit holding the waste is likely to fill with water unless permanent groundwater displacement actions are taken.

Backfilling pits has been resisted by the mining industry as too costly, primarily because it requires operators to move waste material twice — first to mine it from the pit to recover the minerals, and second to move it back. Nevertheless, companies can design their mine plans so that backfilling of pits is an efficient process that is integrated into the mining and reclamation stages of the mine.

Some companies have used pit backfilling as a technique to prevent water contamination from mining wastes. In the late 1980s, mine regulators discovered that a waste rock pile at Lac Minerals Ltd.'s Richmond Hill mine in South Dakota was generating acid, posing a threat to a downstream fishery.

PHOTO: GARY SPRUNG

Poor waste containment practices at mine sites, as illustrated by this break in a tailings dam at the Peanut Mine in Crested Butte, Colorado, lead to water contamination and the need to undertake expensive water treatment.

In 1994, as part of a state-approved mine-closure plan, Lac Minerals developed a waste capping system to isolate its most reactive wastes within its mine pit. Lac Minerals identified its most reactive waste rock and placed it in the pit first. It then covered the reactive waste with an 18-inch thick, low-permeability layer which the company manufactured by mixing non-reactive waste rock with bentonite (a variety of clay). Lac Minerals added a 4.5-foot layer of waste rock to protect the low-permeability barrier from frost and root damage, and then completed the cap with a 6-inch layer of topsoil. The pit chosen for backfilling was said to be ideal because it was in a "high and dry" location, well above the groundwater level, and it received no natural drainage. According to Thomas Durkin, a South Dakota mining regulator, the results of the capping at Richmond Hill have been "very promising", and may spare the operator the need to resort to perpetual water treatment of acid mine drainage.[11] Durkin reports that reactive waste in the pit has generated no drainage since the backfilling and capping were completed. However, the trial period is still very brief compared with the long duration of the risks of contamination.[12]

Plugging Mine Adits — When underground mines are dug through ore bodies rich in sulfides and heavy metals, the mine tunnels and shafts themselves can become sources of water contamination. A technique that has been used to stop the flow of mining pollution from underground mines is the installation of *adit plugs.* An adit plug is used to flood an underground mine by blocking the flow from shafts and adits (tunnels) that provide drainage outlets for groundwater that flows into the mine and becomes contaminated.[13] Reinforced concrete plugs can be inserted into these mine openings, forcing the groundwater drainage inside the mine to rise and totally submerge the acid-generating sulfide materials on the exposed rock faces inside the mine. Although this process may seem paradoxical, flooding can minimize acid mine drainage by cutting off the flow of oxygen to the sulfidic rock, thus stopping the chemical reaction that creates acid mine drainage.

Adit plugging is a risky approach that doesn't always work, however. Under pressure, plugs have blown out of the adits causing major, sometimes catastrophic, discharges of large masses of polluted mine water. Plugging also may create new pollution problems, such as forcing polluted water to emerge under pressure out of other mine openings and ground fissures at the site. Plugging may not be feasible if the local hydrology is too complex or if there are too many passageways to seal.[14]

While most adit plugging accomplished to date has taken place at Appalachian coal mines, plugging has been practiced at some metal mines. At the Summitville Superfund site in Colorado, a cleanup contractor placed reinforced concrete plugs in two adits at the base of the abandoned underground gold mine workings. Before plugging, one of these adits discharged an average of 200 to 300 gallons per minute of acidic, metal-laden water into the Wightman Fork, a tributary of the Alamosa River.[15] So far, the plugs have dramatically reduced the discharge of water into the Wightman Fork.[16] According to James Hanley, Remedial Project Manager of the Summitville site, it is still too early to tell whether the mine plugging has succeeded in slowing down the process of acid formation.[17] Meanwhile, other mine plugging projects are underway at the Sunnyside gold mine in Colorado and the Pitch gold mine in South Dakota.[18]

Submerging Acidic Wastes — Because submerging acidic mine wastes under water isolates them from oxygen, it blocks the sulfide-oxygen-water reaction that forms sulfuric acid. This effect is the basis for a waste-disposal method called *subaqueous tailings disposal.* With this process, mine tailings are piped to a discharge point under water in a prepared impoundment or a natural body of water, such as an artificial pond, a lake, or the floor of the ocean. This technique has been permitted at several mines in Canada. These include the Island Copper Mine in British Columbia, that deposits its tailings in a Pacific Ocean fjord, and copper, zinc, and lead mines which deposit tailings into Anderson Lake, Manitoba.[19] Several studies on subaqueous disposal indicate that this method can retard the oxidation of sulfide minerals in mining wastes, and therefore prevent the generation of acid.[20]

However, controversy arises when mining companies propose to discharge their tailings into natural water bodies. Although the U.S. Environmental Protection Agency currently prohibits discharge of tailings into U.S. waters and seabeds, the EPA and the former U.S. Bureau of Mines have conducted studies on the feasibility of undersea disposal of mining wastes. The A-J Mine at Juneau, Alaska, was seeking a permit to discharge its tailings on the seabed in the mid-1990s. The long-term impacts to the aquatic environment of these discharges remain unknown.[21] Submerging mine wastes into oceans and lakes could cause harm to aquatic life by smothering bottom- dwelling species and increasing the turbidity and heavy metal concentration of water[22] (see Chapter 6).

Blending Wastes

Another way to prevent acid mine drainage is to blend or combine reactive wastes with alkaline materials such as limestone, lime, soda ash, and caustic soda. While these materials will not stop the oxidation of pyrite, they can neutralize acidity that is generated in the waste, and make any runoff from the waste less harmful to the environment. Blending can also prevent the acid-accelerating conditions that arise when *Thiobacillus ferrooxidans* bacteria proliferate in a moderately acid environment.

- Cleanup operators at the Silver Bow Creek mining Superfund site near Butte, Montana, have carried out several waste blending projects, some with encouraging results. In the last few years, the Atlantic Richfield Company and the state of Montana have used special plows to incorporate lime and limestone into metal-laden, acid-generating tailings. The cleanup operators obtain their lime and limestone from a local lime producer and quarry 40 miles from the mine site.[23] These tailings, which were deposited along streams by late nineteenth- and early twentieth-century mining operations, today threaten Silver Bow Creek and the upper Clark Fork River with acid and metal contamination. According to Scott Brown, the site's Remedial Project Manager, the lime treatment has contributed to an improvement of water quality in the Clark Fork River and Silver Bow Creek. Brown attributes the treatment's success to its encouragement of diverse plant growth on tailings piles, which he says has reduced infiltration of water into the piles and prevented the erosion of acid-generating materials into the two waterways. According to Brown, before the tailings were treated, their high acidity and metal content made them unsuitable for plant growth.[24] This is a use of blending to address a Superfund-site contamination condition which should never have been allowed in the first place, but it demonstrates the effects of blending as a technique.

- Barrick Gold Corp.'s Goldstrike mine in northern Nevada, seeks to prevent acid mine drainage by combining sulfidic waste with non-acid generating waste rock during waste disposal. Barrick does not actually blend the wastes; rather it removes potentially acid-generating waste rock to a special part of its waste rock pile and then surrounds the waste on all sides with non-acid-generating waste rock. The result is a protective "cell" that is designed to buffer acid generated in the cell's core. Time will tell whether this experiment, which does not cut off air or water from the sulfides, succeeds in preventing acid drainage.[25]

While neutralizing materials such as limestone may be readily available, buffering environmentally-risky wastes may require vast amounts of alkaline material. Recent research shows that alkaline materials must be intimately blended with sulfidic waste, not just layered.[26] Thus, a successful blending strategy may require too much material, time, and labor to be cost effective.

Moreover, the long-term effectiveness of waste blending to buffer acid generating wastes is speculative. Operators who blend or combine wastes are in effect undertaking a large-scale experiment in acid-base accounting. Offsetting acid-generating materials with acid-neutralizing materials does not ensure that acid mine drainage will not develop over time in a real mine setting. Factors which are hard to control, such as the role of bacteria, and the different reaction rates of sulfide and neutralizing materials, can upset the most hopeful predictions based on a simple surplus of acid-neutralizing materials in waste samples. Therefore, mine operators should be cautious about using waste blending as a primary acid mine drainage control strategy.

Treating Contaminated Mine Wastewater

Preventing water contamination before it starts is always the preferred choice. Existing water contamination from mining, however, often requires pollution *treatment.* Operators can use active or passive treatment techniques to treat water contamination. Active treatment techniques involve considerable human intervention and equipment, such as adding chemicals to polluted mine effluent. In contrast, passive techniques are low-maintenance, requiring little human involvement and few chemical additives. Passive techniques typically rely on the ability of plants and bacteria to mitigate mining contaminants. While passive techniques have not been adopted widely by the hardrock mining industry, they offer promise because of their potential to provide a low cost alternative to traditional treatment methods.

Operators can treat acid mine drainage actively by adding a neutralizing substance, such as limestone, sodium hydroxide, or fly ash, to the acid effluent released from the mine. This neutralizing material reduces the acidity of the effluent and also may reduce the level of metal pollution. Treating polluted water by this method may restore the water to a satisfactory quality. Water treatment is expensive, however, because the need for treatment is usually long term (sometimes permanent), and the treatment system operator must dispose of large quantities of metal sludge (watery masses of metallic wastes) that are recovered during treatment. Despite these drawbacks, active chemical treatment is now the predominant method of treating water contamination at both operating and inactive mines, including Superfund sites.

A less common active treatment technique involves using bactericides to inhibit iron-oxidizing bacteria like *Thiobacillus ferrooxidans*. *T. ferrooxidans* catalyzes the generation of acidity. Bactericides are typically anionic surfactant chemicals that are contained in common products such as laundry detergents, shampoo, and toothpaste. These chemicals work by destroying the greasy, protective cell walls of *T. ferrooxidans* and other acid-catalyzing bacteria, exposing their interiors to the acid that the bacteria help to generate. Bactericides are biodegradable and can provide nourishment to other bacteria which compete with *T. ferrooxidans*. They are usually mixed with water and then sprayed directly on mine wastes.[27]

While used widely and successfully at coal mines, bactericides have had limited impact on the treatment of acid drainage at metal mines.[28] However, the DeLamar silver mine in Idaho has used bactericide treatment successfully to reduce the costs of treating acidic runoff from reactive waste rock piles. Bactericides also have been used at metal mines in New Zealand and India. In addition to treating acid mine drainage, bactericides can be applied to wastes to suppress acid mine drainage by inhibiting the initial establishment of populations of *T. ferrooxidans* and other acid-promoting bacteria.[29] Bactericides are a short-term tool, however, not a permanent waste treatment solution.

An alternative to chemical treatment is *bioremediation*, a set of passive treatment techniques which employ bacterial or other organic agents. One of the most successful bioremediation techniques is the construction of man-made wetlands to route mine effluents through areas stocked with aquatic plants such as cattails. These plants can trap or absorb metals or serve as a growing base for bacteria that function as metal collectors. Near the water's surface, some bacteria derive energy by combining oxygen with iron (a source of acid mine drainage), oxidizing the material to low-solubility iron oxide and causing it to settle out of the water, or "precipitate." Other species of bacteria used in this cleanup method thrive in the anaerobic (oxygen-free) environment found at the base of some plants, where the bacteria obtain energy by reducing sulfates to hydrogen sulfide gas. Hydrogen sulfide reacts with the metals to produce insoluble metal sulfides, which precipitate out and can be recovered and disposed of. This process also consumes acidity, raising pH levels.[30]

The use of wetlands to reduce acid mine drainage has been widespread in the coal industry but, thus far, wetlands have been used only on a pilot scale in metal mining. Richard L. P. Kleinmann, a government expert on acid mine drainage, reports that, although some tests have been successful, it is too early to know the method's usefulness in treating water contamination from hardrock mining. "Ironically, wetlands may trap so many metals that they may create new toxic waste sites. Also, mine effluent in Western mines often contains higher levels of metals such as copper, zinc, and arsenic, that are environmentally more harmful than other metals," Kleinmann says. He also notes that the upkeep of wetlands is often difficult in the dry areas common to many metal mines. Additionally, wetland plants that accumulate toxic metals could pose a health threat to foraging animals. Kleinmann concludes that the use of constructed wetlands at metal mines probably would be most successful in cases where dissolved metal levels in contaminated effluent are low.[31]

Research and experimentation continues on other bioremediation techniques. One method involves incubating waste-metabolizing bacteria in metal drums or tanks, known as bioreactors, and running polluted mine runoff through the tanks. Pilot tests have demonstrated that anaerobic bacteria could be used to recover metal sulfide contaminants from mine effluent. Because metal sulfides precipitate or settle out of water at different pH levels, researchers were able to precipitate metal sulfides (such as zinc sulfide and copper sulfide) selectively by controlling the pH of the mine effluent. This method offers the prospect that metals can be recovered from the sludge produced by some water treatment processes, and the recovered metals can then be marketed commercially.[32]

The need for developing new cyanide treatment processes has become more pressing because many chemical leaching mine sites, where cyanide has been used for a decade or more, now have reached the end of their productive lives. Great quantities of wastes in retired ore heaps will have to be treated and detoxified.[33] A current detoxification method involves rinsing the cyanide-laced heap with water and chemically treating the collected water with hydrogen peroxide, sulfur dioxide, and chlorine to oxidize the cyanide and convert it to a less toxic form. However, chemical treatment can be costly, it may produce toxic byproducts, and it will require careful control of pH levels in the rinse water so as not to accelerate metal leaching.[34]

Bioremediation of cyanide, a less developed technology, may overcome some of the problems associated with conventional chemical treatment. Bioremediation taps the capacity of certain strains of bacteria to break down cyanide to obtain energy. Except for bacterial nutrients, this process does not require additional chemical additives. It also produces few harmful byproducts. Homestake Mining, at a gold mine in Lead, South Dakota, has been using bacteria for more than a decade to oxidize cyanide and convert it to carbonate and ammonia. Another company, USMX, successfully field-tested a bioremediation process at its Green Springs mine near Ely, Nevada.[35] Recent cleanup work by the Environmental Protection Agency at the Summitville Mine Superfund site in Colorado also includes small-scale bioremediation of cyanide leach heaps.[36]

Mine Reclamation

Preventing water contamination from mining requires that careful action be taken both while mining is under way and during the restoration of a site after mining is completed. The techniques for pre-mining planning, waste containment, and treatment of contamination, should also be applied to mine reclamation — restoring land to a productive use after mining. The public commonly recognizes reclamation as a way to make mine-scarred landscapes visually attractive. But reclamation is also a key element in preventing water contamination and sedimentation and in restoring wildlife habitat. The sooner reclamation starts after mining begins, the more effective it will be, because mine wastes will be exposed to the elements for shorter periods of time. In fact, reclamation should be concurrent with ongoing mining operations to prevent post-mining water contamination that will require expensive emergency treatment.

PHOTO: MINERAL POLICY CENTER

This grass-carpeted slope helps hide the evidence of recent mining at the McLaughlin Mine site in California.

As noted above, capping and revegetating waste rock and tailings as soon as feasible after mining can prevent acid mine drainage by reducing water and oxygen contact with wastes; locating waste piles away from water drainage areas works toward the same objective. Also, as waste rock and tailings accumulate, they can be blended with alkaline material such as lime to control potential acid mine drainage. Effective waste blending, however, requires accurate knowledge of the acid-forming potential of wastes.[37]

Proper shaping and revegetation of waste impoundments is important, too. Waste rock and tailings piles should be regraded and contoured into gentle slopes, because steep-sided waste piles are particularly vulnerable to erosion. Topsoil removed during the course of mining must be conserved and protected so that it can later be used to revegetate the site, including capped waste structures. The land should be reseeded with native species of grasses, shrubs, and trees. Because native species are better adapted than imported species to local climate and soil conditions, they will establish themselves more effectively on reclaimed land.[38] Establishing a covering of plants on mined land can be challenging; mine wastes are often toxic to plants because of high acidity and metals concentrations, or unsuitable for vegetation because of the lack of soil nutrients. The operator can create a more favorable medium for growth by adding fertilizers or amendments, like lime, to wastes and soil. Lime treatments can render soils less toxic to plants by lowering acidity and immobilizing metals.[39]

A mine operator must continually monitor surface water and groundwater quality in the vicinity of the mine to assess the effectiveness of ongoing reclamation. Baseline studies of area water quality conducted before mining begins can provide a benchmark for measuring progress in achieving reclamation goals.[40] Kinetic testing to predict acid mine drainage also should be conducted before mining starts and continued periodically during the operation. The knowledge gained in kinetic testing and other sound monitoring procedures can result in a better, more effective reclamation process. Finally, successful reclamation will benefit the mine operator financially by making the need for expensive and long-term waste treatment less likely.

Homestake Mining Company's McLaughlin gold mine, in the Coast Range near California's Napa Valley, is one example of a mine which is successfully carrying out a concurrent reclamation plan using the techniques and principles described above. It is doing so while running a profitable operation. Although McLaughlin's reclamation plan is not typical practice for the hardrock mining industry, it does provide an achievable, cost-effective model for other operations to follow.

Before beginning operations in 1985, McLaughlin performed extensive baseline water sampling at its site and assessed the area's hydrology. McLaughlin subsequently used this information as a guide in designing its facilities and measuring the impact of its operations on the area's water quality.

McLaughlin's careful planning allowed it to obtain a "zero-discharge" permit for the mine under California water regulations. It secured the permit by designing a system of water management that would contain, collect, divert, and reuse water without releasing it into the environment. After screening thirty-four possible sites for its tailings impoundment, McLaughlin selected a site that lay over a natural clay deposit which provides isolation from groundwater. Although the tailings dam does not have a liner, the natural clay deposit was tested to have a very low permeability. McLaughlin designed both the mine's tailings impoundment and waste rock piles with overflow ponds to collect excess water from large storms.[41] The tailings dam has enough backup containment to capture the volume of the probable maximum flood (a 1,000-year, 72-hour storm event).[42] The mine collects seepage water from the tailings dam and waste rock and uses the water in its grinding and ore processing operations. To prevent surface water from contacting mine wastes, McLaughlin built diversion ditches around its waste rock piles.[43]

Early in its mine life, McLaughlin found that it had seriously underestimated the potential of its waste rock to generate acid, mostly because it had relied on inaccurate acid-base accounting methods (see Chapter 3, Acid Mine Drainage). More sophisticated kinetic testing revealed much higher acid-generating potential in its waste rock. The mine responded by adopting a program of identifying waste types by their acid-generating and neutralizing potential. Instead of dumping waste randomly, as they had before, the operation now segregates high-risk, high sulfide wastes, encapsulating them in non-acid generating clays and waste rock.[44]

PHOTO: PHILIP M. HOCKER/ MINERAL POLICY CENTER

The mine regrades, recaps, and revegetates waste rock piles as soon as possible when dumping areas are retired. The company reseeds reclaimed land with native grasses and clover to prevent erosion and restore wildlife habitat. As operations and reclamation continue, the company conducts extensive monitoring of water quality around the mine.[45] So far, testing reveals that the area's water quality is at pre-mining levels.[46] Upon the conclusion of mining, Homestake's objective is to make its mine site available as a biological field station to be managed by the University of California. The company also intends to make its collected environmental monitoring and baseline data available for scientific research.[47]

At the McLaughlin Mine, extensive baseline water studies, careful planning, concurrent reclamation, and continuous monitoring provided for an environmentally responsible mining operation.

The McLaughlin Mine's concurrent reclamation program demonstrates that hardrock mines can be both profitable and environmentally-responsible operations. The company's environmental expenditures, have not burdened it financially. The total preconstruction environmental expenditures at McLaughlin were less than 2 percent of the mine's $284 million capital cost.[48] Ray Krauss, McLaughlin Mine's environmental manager, observes that the cost of the mine's ongoing acid mine drainage management system is nominal, at approximately two cents per ton of ore placed in the pile. Krauss notes that, "this cost is relatively insignificant when compared to what the cost might be to construct and operate a water treatment plant to treat the discharge from an unmanaged waste rock pile."[49]

Mining More Efficiently

Mines also can employ planning and technology to design more efficient operations that generate less waste, use fewer harmful processing agents, and conserve water. For example, a mining company may be able to recover valuable metals left in old waste rock and tailings, cutting back on the amount of new waste material it generates, as well as reducing existing mine waste:

- BHP Copper uses this approach at its Pinto Valley mine near Globe, Arizona. With chemical leaching and SX\EW processes (see Chapter 2, Copper Heap and Dump Leaching), the mine recovers copper from old tailings piles created in the early part of the century.[50] The copper grade in this "waste" material is higher than the grade of new ore being produced at the main mine.

- Viceroy Gold Corp.'s Castle Mountain mine in California's Mojave Desert now delivers cyanide to leaching piles by drip method, using small leaks at perforations in tubing, rather than by spraying from sprinklers. The drip method conserves water, a precious resource in the mine's desert location. In addition, this method reduces the amount of cyanide stored in deadly pools that attract birds and other wildlife.[51] Companies using this technique have lowered processing costs.[52]

- Sometimes it is possible to recover a polluting waste substance, and market it as a product. From 1991-1993, at its Superior mine in Arizona, Magma Copper Company employed methods to recover acid-forming pyrite material at the mine and sell it as an industrial chemical product. The company recovered high-grade pyrite by subjecting its copper tailings to an additional flotation circuit (see Chapter 2, Mineral Processing), enabling a 99 percent pure pyrite concentrate to be skimmed off the top of the tailings slurry. The concentrate was then dried and packaged for sale. Because the tailings that had gone through this process contained less pyrite, they posed a smaller risk of generating acid mine drainage.[53] The company limited its use of this pyrite recovery process, using it only when the prevailing market demand for pyrite materials made it profitable.[54] Unfortunately, according to an official of BHP Copper (the current mine owner), the Superior Mine suspended the pyrite removal process because of limited market demand for its product.[55]

These frugal mining techniques offer economic as well as environmental benefits which the public — and mining companies — can reap. Mining companies can adopt these methods to produce minerals at lower cost, while generating less waste. And by curbing waste generation, mines can reduce cleanup costs and potential legal liabilities for water contamination.

Adopting a New Management Mindset

Adopting environmentally-clean mining techniques alone will not prevent the industry's contamination of water. Companies also must take a leadership role, and instill an uncompromising commitment to environmental excellence in all personnel — from corporate officers to on-the-ground operators. Mining industry management must develop new workforce policies that reward superior environmental performance. For example, management should develop hiring and promotion policies which rate workers as much on their achievement of environmental goals as on their advancement of corporate profitability.

The need for new management policies to broaden the industry's mindset is demonstrated by Echo Bay Corp's slow reaction to repeated incidents of bird and wildlife deaths at its McCoy/Cove mine in northern Nevada. In the late 1980s and in 1990, the McCoy/Cove site's gold and silver cyanide milling operations poisoned more than 1,000 migratory birds with cyanide.[56] In October 1990, the U.S. District Court in Reno fined Echo Bay $250,000,[57] then the largest fine ever issued under the Migratory Bird Act.

The massive bird deaths at McCoy/Cove were caused by a new mill and tailings pond, which the company started operating in July 1989. Upon start-up, flocks of birds were lured to the pond's wet cyanide-laced tailings, where they ingested the deadly toxin. Although large numbers of bird deaths occurred immediately — 372 in the first month of operation — Echo Bay continued to process ore and discharge tailings with high cyanide concentrations. The Nevada Department of Wildlife had even warned Echo Bay a year earlier that its new facility could cause bird deaths if cyanide levels in tailings exceeded a level of 50 parts per million.[58]

For months, as bird deaths mounted, the company neglected to solve the problem. Only in May of 1990 did Echo Bay submit an amendment to its operating plan to install an "INCO/SO_2" cyanide destruction process, after several less expensive attempts to control bird deaths had already failed. When the company installed the new cyanide destruction process in September 1990, bird deaths declined considerably. However, as early as July 1989, in the first month of the mill's operation, Echo Bay had already identified this process as the most effective way to end wildlife deaths. But in a letter to the Nevada Department of Wildlife, the mine's general manager declined to use the process, writing that "due to the expense [of the INCO/SO_2 process] we will actively be pursuing other alternatives."[59]

A greater commitment to environmental excellence by Echo Bay's management could have prevented many of these wildlife deaths. The company's slow response and continued production during the crisis suggested that profitability, rather then commitment to the environment, weighed most heavily in management's concerns. As a local Sierra Club chapter noted during the bird deaths, "Echo Bay apparently decided that the high wildlife deaths were less important than an incremental increase in profit. . . A prudent mining company could have reduced the cyanide levels, taken a somewhat lower profit, and reduced wildlife mortality until the detoxification problems were solved."[60]

Mining companies have the financial resources, technological skills, and human resources to operate mines that prevent water contamination. Yet, as the many cases of water contamination in this book show, most companies do not practice clean mining consistently. And when they do, it is often only at a single "showcase" mine, like McLaughlin, among their many operations. The mining industry's only intermittent embrace of clean mining shows that a real commitment to the environment, beyond the rhetorical, is still lacking in most mining company boardrooms.

We know that the techniques to prevent water contamination from mining exist. Now, however, clean mining must also be sustained by an industry management philosophy that accords protection of the environment the same respect as corporate profits.

New Public Policy Standards

We cannot depend on the consciences of mining industry officials to ensure protection of our water resources. After all, as Samuel Johnson once said, "Conscience is that little voice that tells us someone may be watching." We must adopt laws and policies that require companies to use the cleanest mining methods and that require mines, not the public, to pay for the costs of long-term water cleanup after they end their operations. The following is a set of initiatives that federal and state policy-makers should implement:

Prohibiting Mining in Some Areas

Today's environmental technology and mining methods do have limits; there are situations in which mining activity presents too great a risk of environmental damage. Therefore, we should reserve for the government the authority, or "discretion," to prohibit mining in such cases.

A recent example where governmental discretion was needed is the proposed New World Mine in Montana, near Yellowstone National Park. This mine was proposed to be located in a highly acidic ore body (containing 15 to 30 percent sulfide concentrations), in a severe climate that annually produces 30 feet of snow and has only 23 frost-free days a year. The region is the second most seismically active area in the lower 48 states. The estimated 5.5 million tons of acidic wastes to be generated from the mine would threaten the pure waters of Yellowstone National Park.

The mining company, Crown Butte Resources, proposed elaborate plans to unearth the ore and store wastes permanently at the site, in what it said would be a safe manner. However, many community residents and elected officials did not accept the company's explanations. The environmental risks the mine posed were simply too great. In August 1996, President Clinton, acknowledging the risks, announced that his Administration had struck an agreement with Crown Butte on a plan to pay the company to drop its plans to mine.

The New World deal was a unique exception to the rule. Mines are virtually never prohibited from opening due to environmental risk assessments. Nor do regulatory agencies often have clear regulatory authority to take such steps. To stop the New World Mine, the President of the United States had

PHOTO: PHILIP M. HOCKER/ LIGHTHAWK

The New World Mine, proposed for Henderson Mountain (center) in Montana, demonstrates the importance of providing the federal government with the discretion to prohibit mining in environmentally sensitive areas.

to get involved. There are other unique or fragile areas that don't enjoy the notoriety of Yellowstone, there are other mines that have a high potential for generating water pollution, and current technology and regulatory controls cannot always assure that mining pollution will be averted. We must adopt workable mechanisms for independently assessing these risks and we must prohibit mining where necessary. As a start, the 1872 Mining Law (which allowed the Yellowstone miners to get their "right to mine") must be reformed to reserve discretion for disapprovals.

Conducting More Research and Development of Methods to Prevent Water Pollution

We need more research and development into mining pollution prediction, pollution prevention, and cleanup techniques. Specifically, research is needed to develop ways to make mines produce less, not more, mining waste, and to reduce the amount of water wasted and displaced by mines. To date, most mining industry research has focused on ways to increase mineral production, methods which produce more mine waste and consume more water. This approach must be reversed.

Market-oriented mechanisms, such as assessing pollution taxes on the amount of waste material generated and water used, could serve as strong incentives for companies to reduce their waste. Conversely, tax breaks to mining companies that develop research projects or pilot programs to reduce mine waste and water consumption might serve as positive incentives.

Another means of promoting the development of mining technologies which are less polluting is to require mining companies to report to the public the amount of toxic materials they release into the environment. This approach already has been applied to many other U.S. industries with positive results. Since 1986, the EPA has required many manufacturing industries (but not the mining industry) to report the amounts of certain toxic pollutants they release into the environment each year. The program is called the Toxics Release Inventory (TRI) and is part of the Emergency Planning and Community Right-to-Know Act. Since the TRI reporting began, there has been a marked decrease in the level of pollutants many companies have reported releasing — showing that embarrassed polluters quickly found ways to reduce their pollution loads.

The mining industry's exemption from TRI has recently ended. On 28 April 1997, the EPA expanded TRI to include seven more industry groups. As a result of the expansion, metal mining companies will be required to report toxic chemicals from mineral processing and extraction activities.

Extension of the TRI reporting concept also could lead to mining industry advances in the area of water conservation. Requiring mining companies to report to the public annually on the massive amounts of water they consume and displace for production and processing activities could encourage more efficient use of water by the industry.

Overhauling the Regulatory System

Leaving it up to the mining industry to safeguard water resources from mining impacts has not worked. Reform of the Nation's piecemeal regulatory approach to mining is essential to compel mining companies to consistently apply sound environmental practices. Without effective reform, the mining industry will continue to operate under the weak, incomplete regulatory regime that has demonstrated repeatedly its inadequacy to protect water.

Since 1977, the coal mining industry has been governed by a comprehensive and powerful federal regulatory law, the Surface Mining Control and Reclamation Act. That law, though imperfect, has greatly improved the environment in coal-mining districts. Hardrock mining, however, is not covered by the Surface Mining law. Chapter 8 details many of the weaknesses of the present state and federal hardrock regulatory system.

Overhaul of the regulatory process must start with the establishment of a set of comprehensive, national environmental protection standards that apply to each specific mining activity that can affect surface and groundwater. The standards must be detailed and clear to avoid vagueness and weak application. While the provisions can be flexible, to reflect certain factors

PHOTO: MARTIN KLEINSORGE

Water flows from old mine shafts and across mine waste in the historic Cripple Creek Mining District in central Colorado. To halt continued water pollution from mining, Congress must establish a nationwide cleanup program for abandoned hardrock mines.

unique to certain regions, they must establish national minimum environmental protection standards for water protection. Overall, a sound regulatory program for mining must include:

- Requiring a complete assessment of surface and groundwater hydrology prior to mining.
- Requiring that all mining operations, before and during mining, characterize the potential of their ore and waste rock to generate acid mine drainage. Operators should be required to use both static and kinetic testing to make this determination.
- Establishing a workable process to assess and designate sensitive areas as unsuitable for mining.
- Establishing specific contamination standards for mining pollutants.
- Requiring pollution prevention and pollution containment techniques in all phases of mine operation.
- Establishing comprehensive standards and definitions for mine reclamation activities.

- Requiring mine operators to meet a stringent "best available technology and practices" standard of environmental performance in all their operations.
- Requiring post-mining water quality monitoring to ensure that acid mine drainage does not develop over time.
- Assigning long-term liability for water quality at the mine site to mine operators.
- Requiring mine operators, prior to mining, to post reclamation bonds that cover the potential cost of long-term treatment of acid mine drainage and other pollution.
- Requiring regulators to inspect mines frequently, especially water-related facilities at mines, and requiring mandatory citation of all violations of law.
- Providing the public with full rights to participate in the mining regulatory process, including the right to compel regulatory enforcement by citizen suits and the right to accompany mine inspectors.
- Prohibiting mining operators who have not cleaned up environmental problems and violations at their other mines from obtaining any new mining permits.
- Establishing a funded *Hardrock Abandoned Mine Reclamation Program* to clean up pollution and safety hazards from the more than 557,000 abandoned hardrock mines nationwide.

Now is the Time

The hardrock mining industry has the financial and technological resources to operate mines that do not harm water resources. The industry's literature is now replete with commitments to environmentally-responsible mining. However, the industry has failed to convert this capacity and rhetorical commitment into the daily practice of clean mining. State and federal regulators have been less than diligent in demanding the industry's best environmental performance. As a result, the nation's water continues to be contaminated by mining.

The mining industry can protect our water resources while producing minerals profitably. The techniques to do this are simple and straightforward:

The mining industry can plan carefully before mining to design mines which will not pollute water. It can make sure that waste piles are located properly, and that they are lined and capped adequately to cut off sources of pollution. It can use a variety of known methods to treat and neutralize acid mine drainage and metal leaching. It can carry out reclamation efforts that are contemporaneous with mining to minimize the amount of mine waste material exposed to the elements. It can reclaim mined lands so that they are productive after mining has ceased and do not cause long-term water quality problems. It can practice more efficient mining, to reduce waste and water use. It can develop management policies that instill a commitment to environmental responsibility in every officer and employee. The mining industry also can afford not to mine some areas where high environmental risks far outstrip their capabilities to prevent degradation. In short, the mining industry can protect the nation's water from the pollution its mines create.

State and federal governments also must do more to safeguard the public's water resources from mining pollution. Regulators can protect water resources by adopting and enforcing comprehensive, national, environmental standards and regulations for mining operations. Weak standards and inadequate agency funding must not be allowed to continue to undermine this process. Regulators must undertake timely and effective inspection and enforcement actions on current and closed mining operations. And, like mining companies, regulators must accept that they are ultimately responsible and accountable to the public for any pollution problems that they allow to occur at mines on their watch.

Neither the mining industry nor regulators are likely to reverse their neglectful ways until they are forced. The public must demand change in both the practice and regulation of hardrock mining.

Water is essential to life on earth. Fresh, clean water is precious and limited. Modern civilization also needs supplies of metals — but we don't have to choose between the two. With care and vigilance, we can have both, if we reform the mining industry's careless practices and the government agencies' lax approaches.

♦ ♦ ♦

Notes

1. Lorraine Filipek et al., *Control Technologies for ARD,* MINING ENVIRONMENTAL MANAGEMENT, December, 1996 at 5. [hereinafter *Control Technologies for ARD*].

2. KENNECOTT UTAH COPPER CORPORATION, PROACTIVE, ACCELERATED ENVIRONMENTAL CLEANUP PROGRAM FOR KENNECOTT'S UTAH COPPER PROPERTY (April, 1995).

3. JAMES MCELFISH ET AL., ENVIRONMENTAL LAW INSTITUTE, HARD ROCK MINING: STATE APPROACHES TO ENVIRONMENTAL PROTECTION (1996).

4. MINE WASTE MANAGEMENT, pp.334-354 (California Mining Association, Ian P.G. Hutchison & Richard D. Ellison, eds., 1992) [hereinafter MINE WASTE MANAGEMENT].

5. Clint Strachan & Dirk van Zyl, *Leach Pads and Liners, in* INTRODUCTION TO EVALUATION, DESIGN AND OPERATION OF PRECIOUS METAL HEAP LEACHING PROJECTS, p.195 (Dirk van Zyl, et al. eds., Society of Mining Engineers 1988).

6. GUNDLE LINING SYSTEMS INC., PRODUCT INFORMATION (August, 1988).

7. Strachan & van Zyl, *supra* note 5, p.196.

8. *Control Technologies for ARD, supra* note 1, p.5.

9. MINE WASTE MANAGEMENT, *supra* note 4, pp.344-346.

10. U.S. BUREAU OF LAND MANAGEMENT & STATE OF MONTANA, DEPARTMENT OF ENVIRONMENTAL QUALITY, VOLUME I, FINAL ENVIRONMENTAL IMPACT STATEMENT, ZORTMAN AND LANDUSKY MINES, RECLAMATION PLAN MODIFICATIONS AND MINE LIFE EXTENSIONS, pp.2-58 to 2-62 (March, 1996).

11. *Id.,* p.1-10.

12. Telephone communication with Thomas V. Durkin, Hydrologist, Minerals and Mining Program, South Dakota Department of Environment and Natural Resources (3 February 1997).

13. Richard L.P. Kleinmann, *Acid Mine Drainage in the United States, in* PROCEEDINGS FROM THE FIRST MIDWESTERN REGIONAL RECLAMATION CONFERENCE, CARBONDALE, ILLINOIS, pp.1-6 (1990).

14. Telephone communication with James Hanley, Environmental Protection Agency Remedial Project Manager, Summitville Superfund site (Colorado) (19 July 1995).

15. Raj Devarajan et al., *Interim Project Report, Reynolds Adit Control Program, in* PROCEEDINGS: SUMMITVILLE FORUM '95, p.128 (Colorado Geological Survey 1995).

16. *Id.,* p.132.

17. Telephone communication with James Hanley, Remedial Project Manager, Summitville Mine Superfund site, Colorado (12 February 1997).

18. Memorandum from Allen Sorensen, Colorado Division of Minerals and Geology, to John Constan, Mineral Policy Center (25 June 1996).

NOTES

19. U.S. ENVIRONMENTAL PROTECTION AGENCY, SUBAQUEOUS DISPOSAL OF MINE TAILINGS: A LITERATURE REVIEW, p.5 (July, 1995).

20. Kim A. Lapakko, *Subaqueous Disposal of Mine Waste: Laboratory Investigation, in* PROCEEDINGS OF INTERNATIONAL LAND RECLAMATION AND MINE DRAINAGE CONFERENCE AND THIRD INTERNATIONAL CONFERENCE ON THE ABATEMENT OF ACIDIC DRAINAGE, p.270 (24-29 April 1994).

21. *Id.*, pp.11-27.

22. *Control Technologies for ARD, supra* note 1, p.5.

23. Telephone communication with Russ Forba, Environmental Protection Agency, Remedial Project Manager, Milltown Reservoir Sediments Superfund site, Montana (14 February 1997).

24. Telephone communication with Scott Brown, Environmental Protection Agency, Remedial Project Manager, Silver Bow Creek Superfund site, Montana (11 February 1997).

25. Telephone communication with Nick Rieger, Elko District Office, U.S. Bureau of Land Management (9 February 1997).

26. *Control Technologies for ARD, supra* note 1, p.7.

27. Telephone communication with Vijay Rastogi, President, MVTechnologies, Inc. (9 February 1997).

28. Telephone communication with Robert L.P. Kleinmann, U.S. Department of Energy (3 February 1997).

29. Telephone communication with Vijay Rastogi, *supra* note 27 (9 February 1997).

30. R. L. P. Kleinmann et al., *Biological Treatment of Mine Water — An Overview, in* PROCEEDINGS OF THE SECOND INTERNATIONAL CONFERENCE ON THE ABATEMENT OF ACID DRAINAGE, 16-18 SEPTEMBER 1991, VOL. IV, pp.31-32 (1991).

31. Telephone communication with Richard L. P. Kleinmann, Research Supervisor, U.S. Bureau of Mines (6 July 1995).

32. R. W. Hammack, D. H. Dvorak, & H. M. Edenborn, *The Use of Biogenic Hydrogen Sulfide to Selectively Recover Copper and Zinc From Severely Contaminated Mine Drainage, in* BIOHYDROMETALLURGICAL TECHNOLOGIES, VOLUME I, pp.631-638 (A.E. Torma, J.E. Wey & V.I. Lakshmanan eds., 1991).

33. Telephone communication with Richard Lien, Salt Lake City Research Office, U.S. Bureau of Mines (12 March 1995).

34. SANDRA L. MCGILL & PAUL G. COMBA, A REVIEW OF EXISTING CYANIDE DESTRUCTION PRACTICES, pp.2-14 (1990).

35. R. H. Lien & P. B. Altringer, Case Study: *Bacterial Cyanide Detoxification During Closure of the Green Springs Gold Heap Leach Operation, in* BIOHYDROMETALLURGICAL TECHNOLOGIES, pp.219-227 (A. E. Torma, M. L. Apel & C. L. Brierley, eds., 1993).

36. Telephone communication with James Hanley, *supra* note 14 (19 July 1995).

NOTES

37. MINE WASTE MANAGEMENT, *supra* note 4, pp.410-415.

38. MINE WASTE MANAGEMENT, *supra* note 4, pp.419-421

39. *Control Technologies for ARD, supra* note 1, p.7.

40. MINE WASTE MANAGEMENT, *supra* note 4, pp.505-507.

41. Raymond E. Krauss, *The McLaughlin Mine: A 21st Century Model, in* THE CHILES AWARD PAPERS, p.4 (December, 1993).

42. ECHO BAY MINES ET AL., ENVIRONMENTAL SUCCESS STORIES (1994).

43. Krauss, *supra* note 41, pp. 4-5.

44. Krauss, *supra* note 41, p.5.

45. Krauss, *supra* note 41, pp. 4-5.

46. Telephone communication with Ray Krauss, Environmental Manager, McLaughlin Mine (29 November 1995).

47. Krauss, *supra* note 41, p.6.

48. Raymond E. Krauss, *Reclamation Practices at Homestake's McLaughlin Mine,* MINING ENGINEERING (November, 1990).

49. U.S. ENVIRONMENTAL PROTECTION AGENCY, OFFICE OF SOLID WASTE, THE DESIGN AND OPERATION OF WASTE ROCK PILES AT NONCOAL MINES, p.45 (July, 1995).

50. U.S. ENVIRONMENTAL PROTECTION AGENCY, PROFILE OF THE METAL MINING INDUSTRY, p.63 (September, 1995).

51. ECHO BAY MINES ET AL., *supra* note 42.

52. *Id.*

53. U.S. ENVIRONMENTAL PROTECTION AGENCY, *supra* note 50, p.62 (September, 1995).

54. *Id.*

55. Telephone communication with Jeff Parker, Environmental Manager, BHP Copper (3 December 1996).

56. HANK LESINSKI, ECHO BAY MINES, A CASE HISTORY OF CYANIDE TAILINGS DETOXIFICATION AND MIGRATORY BIRD MORTALITY AT ECHO BAY'S MCCOY/COVE OPERATIONS (received by Mineral Policy Center 9 September 1991).

57. *Miners Dig Deep to Give Birds a Safer Passage,* FINANCIAL TIMES (LONDON), 19 December 1990.

58. Letter from William Molini, Director, Nevada Department of Wildlife, to John B. Walker, Nevada State Clearinghouse, Office of Community Relations (March, 1988).

59. Letter from David Naccarati, Echo Bay General Manager, to Nevada Department of Wildlife (18 July 1989).

60. TOIYABE CHAPTER, SIERRA CLUB (RENO, NEVADA), PRESS RELEASE (28 November 1989).

♦ ♦ ♦

VATERSHED

BOTH ACTIVE AND ABANDONED MINES CAUSE POLLUTION

PHOTO: TOM ZUCCARENO, LIGHTHAWK

There are 557,000 abandoned hardrock mines nationwide. Acid drainage from old mining sites can pollute water for thousands of years after operations have ceased.

PHOTO: BOB OLSGARD

At modern active mines, pollutants can leak into groundwater and nearby lakes and streams. The Flambeau Mine, shown here, was built a mere 140 feet from the Flambeau River in Wisconsin.

MINING CAUSES WATER POLLUTION IN COUNTRIES AROUND THE WORLD

PHOTO: PHILIP M. HOCKER/ MINERAL POLICY CENTER

The tailings dam at the Omai Mine in Guyana ruptured in 1995, spilling 2.6 billion liters of wastewater into the country's main river.

PHOTO: ANN MAEST

Goat crosses a tenuous bridge over thick mining sludge, Rio Locumba, Peru.

There Is No Reason for This Damage to Continue

Careful planning, responsible operation, active community participation, and stringent enforceable environmental standards are essential components of a comprehensive strategy to limit the negative impact of mining on the environment, including our water resources.

This reclaimed waste rock pile, at the Marigold Mine in Nevada, has been recontoured, covered with topsoil, and seeded with grass.

Glenn Miller, an environmental expert, tests the liner material at the Mesquite Gold Mine in Imperial County, California.

PHOTO (LEFT): ANN MAEST PHOTO (RIGHT): PHILIP M. HOCKER/MINERAL POLICY CENTER

PHOTO: WILL PATRIC, MINERAL POLICY CENTER

This beautiful watershed in the Okanogan Highlands region of north-central Washington is threatened by a proposed cyanide-process open-pit gold mine. Local activists and citizens' groups are working to address the potential impacts of this mine.

Mineral Policy Center

Mineral Policy Center is a non-profit organization dedicated to solving the environmental problems caused by irresponsible mining and mineral development. The Center was founded in 1988 by its President, Philip Hocker and former Secretary of the Interior, Stewart L. Udall.

Mineral Policy Center's programs and activities include mining-related research, public outreach, advocacy of legislative and regulatory reform, technical assistance to community groups, and community organizing. The Center's staff work with many independent local and regional citizens' groups concerned with mining projects in America and around the world.

Mineral Policy Center relies upon membership support to implement its programs and activities. With your membership you will receive a year's subscription to *Clementine*, Mineral Policy Center's journal. You will also obtain reduced prices on the Center's reports and publications, and regular updates on mining reform events. Please become a member and help us in our efforts to clean up the mining industry.

About the Book

GOLDEN DREAMS, POISONED STREAMS is a new report from Mineral Policy Center. This book assesses the destructive impact hardrock mining has on water resources, provides an expert scientific and legal framework that concerned citizens can use to understand the problem, and recommends what can be done to stop mining pollution. This book is a vivid factual account of how mining affects water quality. It should be read by anyone who cares about the health and future of our natural environment.

For your copy of this important report, please send a check or money order for $24.95 payable to Mineral Policy Center to:

GOLDEN DREAMS, POISONED STREAMS
Mineral Policy Center
1612 K Street, NW, Suite 808
Washington, D.C. 20006
202–887–1872

FRONT COVER PHOTOS: SPENCER SWANGER, TOM STACK & ASSOCIATES; TOM ZUCCARENO, LIGHTHAWK

NONPROFIT ORGANIZATION
U.S. POSTAGE
PAID
WASHINGTON, D.C.
PERMIT NO.3541

Part II
An Expert's Kit to Water Damage from Mining

Part II of *Golden Dreams, Poisoned Streams* is a collection of technical papers which examines the ecological, human health, legal, and global impacts of water damage from mining. Each of the five essays was written by a respected expert in his or her field of study. The papers can be read separately to research a particular problem associated with mining damage, or taken in aggregate to build greater technical knowledge of the issue. In essence, the second half of *Golden Dreams, Poisoned Streams* is a tool kit designed to help readers become citizen experts on the variety of "real-life," tangible problems which pollution from hardrock mining causes.

To further hone your expert knowledge about environmental problems from mining, please refer to the Bibliography of Suggested Reading.

MINING AND HYDROLOGY

5

Thomas J. Aley and Wilgus B. Creath

Thomas J. Aley is Director of the Ozark Underground Laboratory in Missouri. A registered professional hydrogeologist and geologist, he has more than 30 years experience in the field of hydrology. Mr. Aley received a Masters of Science degree in Forestry from the University of California, Berkeley.

Wilgus B. Creath is a consultant with more than 40 years professional experience as a geologist. He has directed technical research teams and published numerous reports related to environmental impact assessments on geologic and hydrologic resources. He received a Bachelor of Art in Geology at Washington University, St. Louis, Missouri.

There is an intimate link between mining and hydrology. During the mining of an ore, water is a nuisance; it must be diverted and discarded. During milling and processing, water is necessary and must be obtained and used in large amounts. Thus, mining can result in large quantities of water moved and displaced above and below ground. Throughout all stages of mining and processing, water acquires potentially noxious and toxic compounds which must, or should, be treated before discharge from the mine site. Finally, after the mining ends, water flows through the abandoned mine workings and waste piles where it continues to collect contaminants. The result is often a water contamination legacy that lasts for decades or generations.

No matter how carefully a mine is designed and operated, it will always create some impact on the overall hydrology of the area. Impacts can range from minor to catastrophic; some can be mitigated easily while others cannot. The purpose of this chapter is to help the reader better understand how rock, water, and mining interact, to show how the interactions vary dramatically in nature and magnitude between mining sites, and to give the reader a workable understanding of the nature of the interactions.

Understanding Hydrology

Hydrology is the study of the interaction of water with earth. This chapter will present a basic summary of hydrologic principles — in particular, the flow and movement of water through the surface and groundwater. Water is an essential trigger for acid mine drainage, as well as the main transport mechanism for mining's contaminants. These principles also explain how and why mining operations may result in falling water levels in streams and groundwater.

The basic principles of hydrology can be conceptualized as a circular process which has neither beginning nor end. Precipitation falls on the earth in the form of rain, snow, or hail. Some of this precipitation collects to form streams, lakes, and rivers. This is surface water. The water that infiltrates below the surface of the ground constitutes the groundwater system. Ultimately, most groundwater returns to the surface via springs, groundwater discharges to surface water bodies, and through wells and other points that extract groundwater. The hydrologic cycle is completed when water returns to the atmosphere through evaporation from land and water surfaces and through transpiration from plants (see Diagram).

The groundwater system consists of two primary zones. The saturated zone is the zone where most of the openings within the earth materials (often called pore space) are completely filled with water. If the saturated zone is capable of yielding appreciable amounts of water if intercepted by a well, it is called an aquifer. The unsaturated zone extends from the surface downward to the top of the saturated zone. The top of the saturated zone is called the water table. Most of the pore space in the unsaturated zone is not water-filled. Water movement in the unsaturated zone is more or less vertical since water movement in this zone is controlled by gravity. This is in distinct contrast to water movement within the saturated zone. Water movement through the saturated zone is more or less lateral and is controlled by the physical properties of the geologic units and by hydrologic conditions. Recharge is the natural process by which migrating water (rain, snowmelt, surface or underground drainage) replenishes the capacity of a groundwater zone.

Porosity is a measure of the amount of pore space in a material. Porosity varies dramatically among geologic units. For instance, a deposit of sand and gravel will have greater porosity than will a granite block. It is important to note that, although rocks may be hard, they are not solid. Most rocks contain porosity in various forms which can be observed and measured.

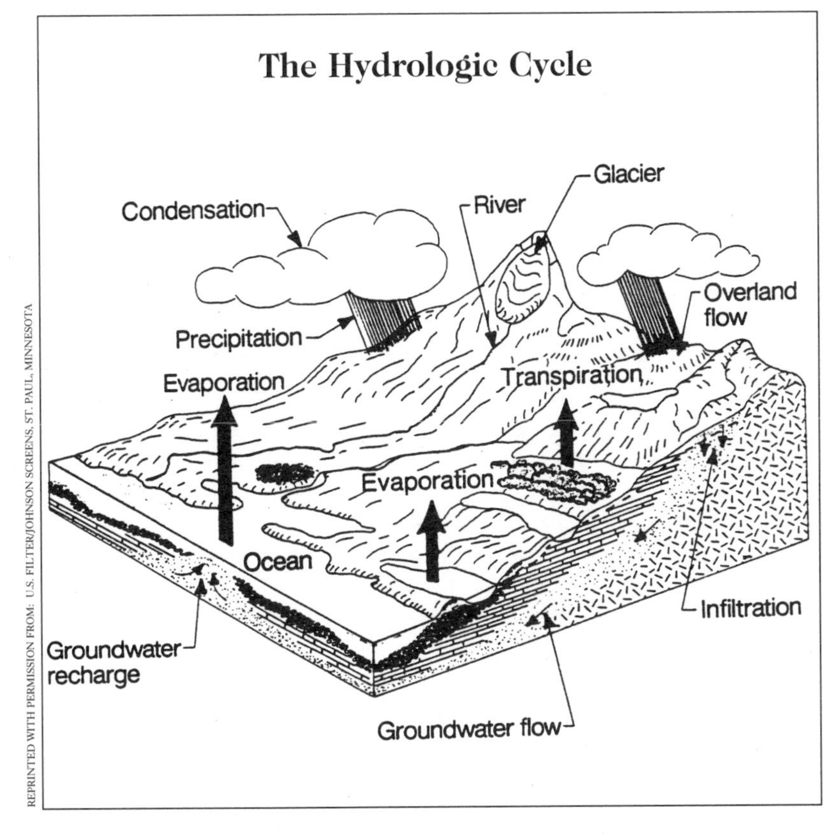

The hydrologic cycle can be conceptualized as a circular process which has neither beginning nor end.

Unfortunately, the concept of a non-solid earth is poorly understood by both the lay public and by many who make land-use decisions.

Permeability is a term which reflects the capacity of a material to transmit a fluid. A highly permeable geologic unit has a greater capacity to transmit a fluid than does a less permeable unit. Hydraulic conductivity is a measurement of the rate at which a saturated geologic unit transmits water. Hydraulic conductivity values in the United States are typically expressed in gallons per day through a cross section of one square foot under standardized groundwater gradient and temperature conditions. Groundwater flow will not occur in an aquifer unless there is a groundwater gradient. On the surface of the land, water flows downhill (which is the same as down-gradient). Flow in an aquifer is also downhill or down-gradient. In the absence of other differences, the rate of flow in an aquifer increases as the gradient increases. Flow directions and groundwater travel rates within portions of an aquifer can be changed by the pumping of wells or mines located within the aquifer.

Hydraulic conductivity values can vary dramatically from one type of earth material to another. The values also can change drastically from place to place within a mass composed of a single type of earth material. Permeability not only differs from one type of rock to another but also can change from place to place within a mass composed of a single rock type. Furthermore, permeability is generally not the same in all directions. In rocks such as sandstones and shales, water moves much more easily parallel to the bedding than it does perpendicular to the bedding. Open fractures can affect permeability dramatically and increase the rate at which water can move through the rock mass. Rocks such as limestone, marble, and dolomite can have vast systems of caves and smaller openings dissolved in them by groundwater. In cave regions, water can move through the rock mass rapidly and for long distances.

As an example, groundwater tracing studies in the Eleven Point and Current River basins of Missouri have demonstrated groundwater travel at rates as fast as 3 miles per day over a 40-mile distance.[1] In contrast, in many other geologic settings, groundwater movement may be only a few feet per year.

In sedimentary rocks and unconsolidated sediments, which comprise many open pit mines, groundwater is contained in, and moves between, the individual grains of the rock or the sediment. In crystalline rocks such as quartz, which comprise many underground mines, most of the groundwater is transmitted through fissures and fractures; very little water moves through the rock body itself. Therefore, when we review the results of a geological study, it is important to know the point of view and the expertise of the investigator. A mining geologist or engineer is interested in the mineral content and competency of the rock itself and may overlook or misinterpret critically important hydrological issues. This has caused many problems in the past.

Now that surface water and groundwater have been differentiated above, it should be understood that surface water and groundwater are intimately connected and, at times, may be considered one and the same. For ease in understanding, visualize some of the water on the surface on Monday percolating through the soil and rock and becoming groundwater on Tuesday. This water may enter a mine on Wednesday, become contaminated with heavy metals on Thursday, and return again to the surface on Friday. This is not an oversimplification. While true in most cases that the "days" in this example may actually be longer in time, it is also true that in some cases our Monday through Friday example spans less than five days.

Hydrology and the Spread of Mining Contaminants

The opening of a mine, whether that mine be surface or subsurface, is similar in many ways to digging into a sealed toxic waste dump. Many mineral deposits are the equivalent of contained hazardous waste sites. Hazardous materials contained by nature are opened by mining. The public would be very reluctant to permit someone to dig into a contained hazardous waste site in search of valuable materials that might be present in the wastes. Even if such an excavation were permitted, great care would be demanded in the entire operation, and effective resealing of the site at the end of the project would be required. The contrast with mining activities is clear. The same

PHOTO: PHILIP M. HOCKER/MINERAL POLICY CENTER

Gypsum stacks, like the one shown here in central Florida, are composed of the waste material produced in phosphate mining. Processing water and gypsum stacks have contaminated underlying aquifers and drinking water wells.

caution for the protection of water resources and public health should be exercised in opening new mines as would be exercised in reopening a closed hazardous waste site.

The mining process contaminates water because mining and milling result in more contact between the minerals and water. This increases the concentrations of contaminants in water. The mining process also increases the amount of water that contacts minerals, therefore increasing the volume of water contaminated. Mining, by exposing vast quantities of encased rock and soil to water and oxygen, sets in motion the formation of acid mine drainage, which mobilizes many environmentally harmful metals.

Mining generates two basic categories of pollutants: physical and chemical. Most physical pollutants do not come from mines but from the processing facilities where the minerals are upgraded or concentrated. Major chemical pollutants are those associated with acid mine drainage from the mines and tailings piles and the escape of various chemicals such as cyanide, hydrochloric acid, and sulfuric acid used in the cleaning and processing of ore. Each type of pollutant is transported by surface and groundwater, often for a considerable distance away from the original source of contamination.

Physical pollutants consist of suspended solids such as sand and silt-sized particles or clays washed from the ores or produced by the milling and concentrating processes. These materials affect water quality by causing the receiving streams to become turbid (less clear, often because of sediment). This can cause adverse environmental consequences such as reduced oxygen levels in water and the smothering of stream bottom habitat. However, troublesome physical pollutants are relatively easy to treat and inexpensive to remove through simple filtering and settling before discharging mine waters into the surface water system.

Chemical pollutants are a more difficult matter. Acid mine drainage may be the most visible form of chemical pollution because of the formation of the slimy, yellow-brown precipitate sometimes called *yellowboy*. However, the leaching of heavy metals into surface streams and groundwater is the most deadly form of chemical pollution and results from water infiltrating into, then discharging from, mines and tailings piles. Acid mine drainage not only carries the heavy metals that are deposited by mining activities themselves but may be the cause of other heavy metals being mobilized. Once an acid stream enters the groundwater system, it can mobilize and spread the additional minerals it contacts in underground rock. Depending on the rock chemistry, the acidic stream can gather new toxins or those toxins already in the stream can precipitate out. Once liberated, heavy metals may travel great distances. Studies conducted by B. A. Kimball (in press) demonstrated that, although a water-treatment plant had removed 75 percent of the toxic metals dissolved in a Colorado mine's discharge waters entering the Arkansas River, the heavy metals attached to colloids (small, suspended particles) were unaffected by treatment and, at the time of the study, had been carried 140 miles downstream. Although colloids settle in the river bed, spring runoff flushes them back into suspension and remobilizes them year after year, providing a continuing source of pollution for the river.[2]

How Mining Alters Hydrology

Mining interrupts the natural surface and groundwater flow patterns in an area. In many areas there are two or more aquifers beneath a particular parcel of land. Such layering of aquifers is common where there are lateral zones of rock which have low hydraulic conductivity. The aquifer on top of such a low hydraulic conductivity zone is commonly called a perched aquifer. Such aquifers can be extremely valuable water supplies for local use; they often account for springs and sometimes are critically important in maintaining perennial flow in surface streams. The construction of mine adits, tunnels, and shafts can drain perched aquifers, causing springs to cease flowing and local wells to "go dry." In many cases, the water from perched aquifers is drained into deeper aquifers which can be tapped only by deeper (and more expensive) wells.

In the case of open-pit mining, the increasing size of the excavation creates an increasingly large snow and rain catchment basin, which funnels greater quantities of water downward through the ore-bearing rocks (thus increasing the acid mine drainage and heavy metal discharge). Mining alters and interrupts surface and groundwater flow by excavation and topographic alteration, the construction of tailings facilities such as dams and ponds, the installation of processing facilities such as leaching pads, by dewatering, and by the tunneling, both vertical and horizontal, associated with underground operations. The surface and subsurface natural drainage change immediately upon the initiation of these alterations, and as mining progresses. As there is a continuous increase in the size of the open pit, in the number of tunnels and shafts, or in the amount of tailings, there is a continuous moving of hydrogeologic boundaries. The characterization of the hydrology of a mine site is never static because of these continuous changes.

At abandoned mines, deep shafts and the presence of tunnels at depth in the same area as the permanent storage of tailings present the possibility of aquifer contamination that otherwise would not occur. At Canon City, Colorado, contaminants consisting of radionuclides such as uranium, radium and thorium, and metals such as cobalt, selenium, molybdenum, and nickel from a uranium processing facility, leaking from unlined containment ponds (and arguably from other ponds with ruptured liners), found their way into the shallow groundwater system. Some are concerned that the presence of the nearby 1,084 feet deep Wolf Park underground coal mine shaft may provide a path for the shallow groundwater contaminants to migrate into the deeper groundwater system. At present, there is no proof that such migration has occurred at this site, but adequate investigations are difficult and expensive.[3]

Dewatering Impacts

Because working levels of most deep mines are below the water table, large quantities of water must be pumped and discharged from the mine continuously. This may cause dewatering of an aquifer over extensive areas and affect the flow of surface streams. Dewatering has different meanings to different persons. To the miner, removing surplus water from the mine site is highly desirable and often essential. To the family that loses its water well or the ability to irrigate its farm as a result of mine dewatering, the activity is certainly less than desirable.

Dewatering by Underground Mining

In underground mining, pumping is used to drain water when groundwater recharge in the unsaturated zone is intersected by mine workings. This can result in a decrease in recharge to deeper aquifers important to adjacent landowners. More extensive pumping typically occurs when the mining is within the saturated zone (and thus below the water table).

In many aquifers, zones of higher permeability are separated by zones of lower permeability. If the mining occurs in a portion of the aquifer with low permeability, the volume of dewatering may not be particularly great.

PHOTO: JEFF WIDEN, MINERAL POLICY CENTER

The open pit at the Bingham Canyon Mine in Salt Lake City, Utah (shown here) measures one-half mile deep and almost 2.5 miles across. Continuous pumping prevents groundwater from filling the pit.

However, if the mining occurs in a portion of the aquifer with higher permeability, then the extent of pumping will be greater.

If a zone of higher permeability overlies a zone of lower permeability, water may be perched or trapped in the zone with the higher permeability. Vertical access and ventilation shafts often pass through such zones in route to deeper rock units where minerals may be located, and these penetrations through the zone of perched groundwater may release the water beyond the zone. Engineers usually try to seal off the more permeable rocks encountered by the shafts — with varying degrees of success. The degree of success almost always will decrease with time. Often these perched zones are important water supplies for local springs and wells. Even after mining ends, the shafts may divert waters permanently from the perched zones into deeper portions of the earth.

Dewatering by Open-Pit Mining

Excavations resulting from open-pit mining also cause a depression in the water table. Once a large excavation is created below the water table, flow will take place to fill this extra space, lowering the water table in the vicinity. Pumping is necessary to keep the inflow from reaching the excavation, which further disrupts groundwater conditions and extends the size of the area with a lowered water table. When the groundwater levels are lowered, flow from area springs may diminish, stream flow may be reduced, and the levels of some lakes and ponds may go down. The yield of water wells may decrease, sometimes to the point of going dry.

The rate of mine dewatering must at least equal the rate at which the water moves toward the mine, whether that mine be open-pit or underground. The size of the area affected depends on the nature of the geologic unit, its permeability, and the presence of localized flow routes through the geologic unit.

Dewatering in Karst Areas

Some of the greatest problems related to dewatering are found in karst areas, since these settings are characterized by dissolved-out networks of connected openings. There have been hundreds of groundwater traces conducted with florescent dyes which have demonstrated that rapid and long-distance water transport occurs in karst areas. It is not uncommon for water to travel through the groundwater system at a rate of thousands of feet per day — sometimes for many miles. In some cases, water sinking in one stream basin may flow underground and discharge from a spring in a different stream basin.

It does not require extensive mining to alter the hydrologic regime in karst areas. In the New Lead Belt of Missouri, a perennial spring used to water livestock on a local farm ceased flowing soon after a large diameter mine ventilation shaft was drilled. Where water once discharged from the spring, air now moved into the same opening in route to the mine. All of the spring water had been diverted into the mine. A similar case reportedly resulted in rapid declines in water levels of the aquifer supplying local wells and necessitated construction of new, and much deeper, wells.[4]

Continuous Dewatering

In the Tri-State (Missouri-Kansas-Oklahoma) lead and zinc mines, the natural level of the water table was usually higher than the mines. Constant pumping was necessary to control flooding in the mines. Decreases in the water table elevation resulting from the pumping were as great as 150 feet, demonstrating the magnitude of dewatering activities. When pumping stopped, the mine drifts and shafts filled with water. Dissolved zinc concentrations now average an incredibly large 9,400 parts per billion in some of these mine waters now discharging to the surface.[5]

Drilling of *adits* (lateral mine passages) into an aquifer intercepts and transmits more water than a vertical shaft or well simply by exposing more area of the aquifer to drainage. Early human water works called *kanats* are evidence that this concept has long been understood. These water-collecting tunnels of the Middle East are the among the greatest water works of the ancients. The oldest kanat is the tunnel of Negoub, built in 800 B.C., to sup-

PHOTO: JEFF WIDEN, MINERAL POLICY CENTER

Pictured here, a gated adit at the Mary Murphy Mine reclamation site in Buena Vista, Colorado.

ply water to Nineveh. The horizontal tunnels connect a series of vertical shafts. Water drains into the lateral tunnels which is then withdrawn through the vertical shafts. The present city of Teheran, Iran, is still underlain by a network of kanats as large as 16 miles long and 500 feet below the surface.[6]

Dewatering in Desert Regions

Nevada is the driest state in the United States. Extensive open-pit mining of low-grade gold deposits is expanding rapidly in Nevada's Humboldt River Basin. The potential hydrologic effects of this mining have received study by the U.S. Geological Survey.[7] While the mines are being developed and operated, pumping of groundwater lowers water levels in extensive areas. The area where groundwater levels have declined as a result of pumping by two mines in the Carlin Trend alone (a broad area of low-grade gold mineralization in northern Nevada) is 350 square miles. This pumped water is discharged to surface streams and is increasing their flow. After mining ceases, water which might otherwise have maintained flows in surface streams instead will enter the groundwater system to replace water extracted by mining.

Groundwater replacement will require years, and probably decades, to reestablish near-natural conditions. This is not an exaggeration. For example, for the Goldstrike Mine in Nevada's Carlin Trend area (see Chapter 3, Water Use, Water Waste and Photo), a computer model predicts that the water table will return to "near normal" a century after pumping stops. Unfortunately, this model is wrong and was based on incorrect engineering assumptions — a projected 1993 pumping rate of 10,330 gallons per minute. In actuality, the pumping rate in 1993 was already averaging 68,000 gallons per minute, a six-fold discrepancy.[8] How many centuries in excess of the original prediction will be required for water table recovery after the Goldstrike mine is exhausted and abandoned? In an arid area, the lowering of groundwater levels by dewatering, coupled with the disruption of surface flows, dries up streams and springs throughout the affected area. This poses significant threats to riparian (riverside) vegetation and aquatic life, and also affects the income of those who use groundwater for irrigation.

Land Subsidence and Collapse

Mine-related subsidence also can affect an area's hydrology adversely. Mine subsidence is the collapse, or shift of, the land mass (including the surface) located over, or adjacent to, an underground mine. Mine subsidence may cause fractures to develop beneath streams, lakes, and ponds leading to the partial or complete loss of water by its drainage into the mine. This can have a serious effect on public water systems and aquatic life, and catastrophic inundation of mine workings may occur. The fractures resulting from subsidence of underground mine workings may lower groundwater levels and decrease the groundwater supply. There may be an increase in streamflow rates for gaining streams because of faster movement through the fractured strata, or there may be a decrease in streamflow rates because of water diversion through the fractures.[9]

A good example of mine subsidence affecting an area's hydrology occurred in 1989 when a portion of the Atlantic Cable underground mine, located immediately under Silver Creek in Dolores County, Colorado, caved to the surface.[10] A whirlpool formed in the creek above the breach in the mine roof. The water from the creek surged into the mine and then drained out through a long drainage adit that flowed out of the mine — down gradient from the stream. The drainage caused minor flooding and seepage problems for several residents. In this case, the problem was not catastrophic and was corrected; however, other collapse problems are not so easily remedied and are far more serious:

In June of 1994, a 15-story-deep sinkhole opened in an 80 million-ton pile of phosphate and gypsum waste at IMC-Agrico New Wales plant in Florida. The hole is estimated at 2 million cubic feet, large enough to swallow 400 railroad boxcars. The cave-in dumped 4 to 6 million cubic feet (20 million pounds) of toxic chemical material into the Florida aquifer, which provides 90 percent of the state's drinking water. The effluent contained 17 heavy metals or other toxic substances including lead, arsenic, chromium, mercury, and cadmium.[11] Although the company has spent $6.8 million trying to plug the hole and control the spread of contaminants, it is highly unlikely that the pollutants can be recovered or that the damage to the aquifer will be corrected.

Dewatering of alluvial aquifers can result in compaction of the aquifer materials. This occurs because water is a significant component of the aquifer matrix. When water is removed by pumping, the packing arrangement of the aquifer particles changes to physically adjust for the missing water. Aquifer compaction is a one-way process; an aquifer that has been dewatered and compacted has permanently lost some of its water storage capacity. Compaction of the alluvial aquifers also can result in land subsidence. Such subsidence can damage highways, pipelines, and buildings.

Some of the gold mines in the Far West Rand of South Africa are in limestone overlain by deeply weathered material called residuum. Dewatering of the mines reduces the buoyant support of the land surface and has induced numerous catastrophic sinkhole collapses, some of which have occurred beneath homes and resulted in multiple deaths.[12]

Mining, Hydrology and Future Land Use

Mining, milling, and refining activities can create significant limitations for future land use. These problems have been reduced in recent years by increased emphasis on surface land reclamation after mining. However, the underground mines and their man-made connections with the surface can create unanticipated groundwater pollution problems for industries that locate on lands that have been mined. As an illustration, hazardous wastes from a surface industry in southwest Missouri infiltrated into old mine workings and migrated through the old mines to springs and streams in the region. The loss of future land use because of irresponsible mining and milling practices is not confined to the surface. In southern Colorado, the holders of approximately 10,000 acres of coal leases believe they are losing an estimated $10 million in potential royalties, because an adjacent milling operation allegedly disposed of or spilled radioactive wastes into shafts and tunnels of abandoned coal mines.[13] As a result, the mineral owners claim mining companies are reluctant to risk accepting the liability of reopening the abandoned workings although the seams may once again be commercial.

Whether underground or open pit, a major aspect of mining that can adversely affect hydrology and the quality of an area's water resources is tailings disposal. It has been estimated that this country currently has over 2,000 waste piles consisting of approximately 50 billion tons of mining waste — not including displaced overburden — spread over several million acres of land.[14,15] These wastes are leaching a wide range of contaminants, including acid and heavy metals, into the groundwater and adjacent streams.

How a waste disposal site (impoundments, waste piles, etc.) is constructed will affect the flow of groundwater and surface waters. Conversely, the water flow will affect the stability and integrity of the waste disposal site. The site will create a new interaction between water and earth. Most importantly, the interaction of a tailings pile with an area's hydrology increases or diminishes the likelihood of groundwater or surface water contamination taking place. Chapter 4 discusses many appropriate waste control and pollution prevention measures necessary to prevent or minimize hydrologic mining impacts.

Protecting Water Resources from Mining Impacts

There are a number of steps that can be taken to reduce mining's adverse impacts on hydrology. These include:

Better Baseline Studies

Mine operators should conduct competent, detailed baseline studies on hydrology and water quality at proposed mine sites. In the absence of adequate hydrological baseline data, it is not possible to assess a mine's potential impact on the environment and, further, operators may find their companies responsible for restoring in-stream water quality to a level neither historically nor biologically justified.

The Summitville Mine disaster in Colorado (see Chapter 1, The Sorry Legacy) provides some insight into the problems caused by inadequate baseline and other fundamental studies. Summitville is a notorious abandoned heap-leach gold mine located in the San Juan Mountains of Colorado. Galactic Resources Inc. abandoned the mine in 1992, leaving the State of Colorado and the Federal government with a site containing a 10-million ton heap leach pad fully loaded with cyanide, 15 million tons of sulfide-rich waste rock, a mine adit flowing acid mine waste and 550 acres of disturbed earth. The taxpayers have been left with a cleanup bill estimated conservatively to exceed $120 million, an amount more than the gross value of minerals extracted from the mine.

Hydrological baseline data as well as baseline data used to characterize the acid- and toxic-forming character of the waste rock at Summitville were inadequate.[16] Most significantly, the climatological data for Summitville, essential to determining the water balance in heap leach pads, were drawn from precipitation and evaporation rates at other locations and were not site-specific.[17] As a result of not collecting adequate and accurate site-specific climatological data, the water level rose to dangerous levels in the heap. To reduce the water level in the heap, the company discharged the cyanide-tainted water in ditches, ponds, and channels, resulting in waste drainage to Whitman Fork and Cropsy Creek, both tributaries of the Alamosa River.[18] Ironically, the Summitville Mine was permitted as a zero discharge mine, thus avoiding compliance with the federal Clean Water Act's

PHOTO: ANN MAEST

An ecologist conducts water quality sampling in Meadow Creek, which drains the Blackbird Mine in Idaho.

effluent discharge permit requirements (see Chapter 8, The Clean Water Act). The basis of the zero discharge determination was the flawed precipitation data that the company furnished the state.

At the time Summitville was permitted (1984), operators were not required to collect any baseline data. Due to the disaster, the state now requires such data. As of July, 1994, Colorado regulations require that operators collect a minimum of five quarters of surface and subsurface hydrological baseline data, and that they characterize the ore and waste rock that may be disturbed by mining.

The availability of adequate baseline data collected before the mine was permitted to operate would have allowed more realistic engineering for the project and certainly would have helped in decision-making regarding the selection of remedial cleanup actions. The ultimate consequences of the Environmental Protection Agency's emergency actions on the surface and groundwater resources of the area are unknown at this time. Because the cleanup has been undertaken as an emergency response action, there was limited opportunity for either public oversight or public comment through October, 1994.

Better Engineering and Waste Containment

Consistent application of sound environmental engineering and water containment practices throughout all phases of a mining operation is essential. In some difficult pollution cases, less conventional methods of dealing with mining waste might be considered. Developing new mineral processing techniques that involve eliminating chemical processing at the site by shipping a concentrate away for final processing might solve some on-site pollution problems. To the lay reader this proposal may not seem unusual. However, because the ratio of mine waste material to desired mineral is so enormous, it has been almost axiomatic in the industry that processing must be done in the area where the mining takes place. Other options are seldom considered. The junior author of this article is currently planning a small project in Colombia using gravity separation and froth flotation to produce concentrates from a number of mines which will then be transported to a centrally located cyanide facility for final processing. This method, originally conceived because of hydrological and environmental considerations, provides an added bonus of increased security for product and personnel in a high risk area.

More Research

More research needs to be done on how pollutants move and are held within the hydrological system. Without this knowledge, it will be difficult to design mines for protection of water resources, and to treat water that has been polluted by mining operations properly. The need for more research has become compelling since most of the economical metal oxide deposits in the United States are being depleted. At many mines, the oxidized portion of the ore body has been mined and the metals have been recovered by pad-leaching techniques. Large reserves of sulfide metal resources remain, which will be the focus of new sulfide-leach techniques to make these deposits economical. The control of acid mine drainage from sulfide processing and sulfide wastes is even more difficult to achieve than for oxide deposits.[19] Without continuing research, it is possible that future mining may present greater threats to the hydrologic environment than those of mining in the past.

Give More Value to Water as a Mineral Resource

The constant search for water at ever greater distances on behalf of communities and industry demonstrates that water is an exception to the general rule that low-value mineral commodities do not travel very far in trade.[20] In fact, at many sites, the long-term value of the water being protected may well outstrip the short-term value of the minerals being mined. Few think of water as a mineral. In the past we frequently have written off or ignored the value of water in our cost-benefit analyses and short-term uses versus long-term productivity calculations. This attitude is irresponsible and clearly unacceptable. We must balance our long-term needs for water, which is essential for life on earth, with our short-term needs for other, less critical minerals, such as gold. We need both, but typically we cannot maximize the production of both at the same site.

♦ ♦ ♦

Notes

1. This groundwater tracing study was conducted in a region where extensive lead and related mineral exploration is currently being conducted on National Forest land.

2. B.A. Kimball, E. Callender and E.V. Axtmann, in press, *Effects of colloids on metal transport in a river receiving acid mine drainage, upper Arkansas River, Colorado, U.S.A.*, APPLIED GEOCHEMISTRY.

3. Colorado Department of Health and Environment, personal communication between author and staff (1977).

4. Data based on field investigations by Thomas Aley and other incidents reported to the authors.

5. J.H. BARKS, U.S. GEOLOGICAL SURVEY, WATER-RESOURCES INVESTIGATIONS 77-75, EFFECTS OF ABANDONED LEAD AND ZINC MINES AND TAILINGS PILE ON WATER QUALITY IN THE JOPLIN AREA, MISSOURI (1977).

6. P.T. FLAWN, ENVIRONMENTAL GEOLOGY, CONSERVATION, LAND-USE PLANNING, AND RESOURCE MANAGEMENT, HARPER & ROW, NEW YORK, NEW YORK (1970).

7. JAMES E. CROMPTON, U.S. GEOLOGICAL SURVEY, WATER-RESOURCES INVESTIGATIONS REPORT 1994, POTENTIAL HYDROLOGIC EFFECTS OF MINING IN THE HUMBOLDT RIVER BASIN, NORTHERN NEVADA, p.4233 (1995).

8. R. Manning, *Going for the Gold,* AUDUBON, (VOLUME 96, NO. 1), pp.69-74 (1994).

9. M. M. Singh, *Mine Subsidence, in* SME MINING ENGINEERING HANDBOOK, SOCIETY FOR MINING, METALLURGY, AND EXPLORATION, INC. (1992).

10. M. W. DAVIS, COLORADO GEOLOGICAL SURVEY, THE USE OF TRACER DYES FOR THE IDENTIFICATION OF A MINE FLOODING PROBLEM, RICO, DOLORES COUNTY, COLORADO (Open File Report 1991-1992), p.20 (1994).

11. M. Satchell, *Sinkholes and stacks,* U.S. NEWS & WORLD REPORT, (VOLUME 118, NO. 23), pp.53-56 (1995).

12. P.T. Foose, *Sinkhole formation by groundwater withdrawal, Far West Rand, South Africa,* SCIENCE, (VOLUME 157, NO. 3792), pp.1045-1048 (1967).

13. W.B. Creath, personal communication with mineral owners (1991).

14. L.M. Kaas & C.J. Parr, *Environmental Consequences, in* SME MINING ENGINEERING HANDBOOK, SOCIETY FOR MINING, METALLURGY, AND EXPLORATION, INC. (1992).

15. Terri Aaronson, *Problems Underfoot,* ENVIRONMENT, (VOLUME 12, NO. 9), pp.17-29 (1970).

NOTES

16. J.A. Pendleton, *The Summitville gold mine and heap leach, Part One; The Problems,* THE PROFESSIONAL GEOLOGIST, (VOLUME 32, NO. 1), pp.9-10 (1995).

17. L.J. Danielson & Alms, *The Summitville Legacy: Where do we go from here?, in* PROCEEDINGS: SUMMITVILLE FORUM 1995, COLORADO GEOLOGICAL SURVEY, SPECIAL PUBLICATION, (VOLUME 38), p.375 (1995).

18. COLORADO SPRINGS GAZETTE TELEGRAPH, 17 June 1995.

19. T.V. Durkin, *New Technology is needed to manage sulfide mine waste,* MINING ENGINEERING, (VOLUME 47, NO. 6), p.507 (1995).

20. FLAWN, *supra* note 6.

♦ ♦ ♦

Mining Waste Impacts on Stream Ecology

6

Robert P. Mason, Ph.D.

Robert P. Mason is a specialist in the fate, transport, and transformation of trace metals in the environment. He is currently Assistant Professor in the Environmental Chemistry Section at the University of Maryland's Chesapeake Biological Laboratory, Center for Environmental and Estuarine Studies. Dr. Mason has a Ph.D. in Chemistry.

The physical and chemical impacts and the consequences of mining activities on water quality and aquatic life are extensive and environmentally damaging. Mining operations can result in the discharge into streams of high levels of acidity and the release of toxic heavy metals and metalloids, as well as large quantities of sediments. Many of the impacts of these contaminants are lethal and dramatic, such as the sudden killing or disappearance of fish and insect life. Cyanide, a lethal toxin used extensively in gold and silver processing, also poses a danger to stream ecology.

However, some of the most damaging aspects of acid mine drainage and other mining contamination of water are not the dramatic, short-term lethal effects but the long-term, insidious degradation of the stream ecology. Often the primary victims are those species of plants and animals occupying the base of the food chain, such as insects, other invertebrates, and algae. Stream predators farther up the food chain — such as fish and crustacea — depend on the invertebrate community as their main food source. Thus the depletion of this vital community leads to longer-term declines in the health and populations of stream predators. Recent research suggests that terrestrial

PHOTO: ROB BADGER

Poisoned stream flows from the Summitville Mine in Colorado have killed aquatic life along 17 miles of the Alamosa River (shown here).

mammals which live in close association with streams also may suffer adverse health effects from mining contamination.

These forms of mining contamination can persist for long periods of time and affect large areas. For example, heavy metals discharged by mining activities often can be transported hundreds of miles away from their source. These metals then can continue to threaten the health of aquatic life further downstream.

In this chapter we explore the environmental impacts of the contaminants released by mining — the acidity, high heavy metal concentrations, and cyanide — as well as the physical disturbance of streams. The focus will be on lethal, sublethal, and chronic effects of these contaminants on aquatic life. The chapter will discuss these topics in the context of the overall stream community and its interwoven food web. It also will discuss the impacts of mining wastes on the mammals and plants that live in association with streams.

The Ecology of a Stream

When mining wastes enter a watercourse, the effects may be felt not only in a handful of clearly noticeable species but also throughout a broad community of interdependent living organisms. The adverse impacts of hardrock mining can reach deeply and widely throughout this stream community, affecting the health and viability of all the species living in or dependant upon that water body. Thus a true understanding of the magnitude of those impacts must be based on a recognition of the complex world of stream ecology.

Each river or stream exhibits its own unique ecology, and the principal factors determining the structure of stream communities are the rate and seasonal variation of stream flows. In high flow streams, the community is dominated by those organisms able to attach themselves to a suitable substrate to maintain their desired habitat.[1] Thus, the benthic (bottom dwelling) community is dominant and critical in many streams. Variable discharge of water into streams can have an important impact on stream life: Stream sediment may be scoured and eroded during periods of high flow, while low flow conditions can lead to enhanced sedimentation, a buildup of finer sediments, and the potential smothering of the bottom-dwelling organisms.[2]

There is a wide variety of substrates (e.g. sand, silt, stones) available within each stream as a result of the different flow patterns from the stream edge to its middle.[3] This variety allows diverse communities to develop, with each organism finding a suitable niche. The flowing stream also provides a well-mixed environment, and low oxygen is seldom a problem in fast flowing waters.

For many streams, the main source of food is the organic matter from the stream's surrounding environment (e.g., leaf litter). Most streams, therefore, are dominated by herbivores (plant-eaters) and detritivores (animals that consume dead organic matter). In contrast, primary production (i.e., plant or algal photosynthesis) accounts for little of the energy transferred to higher food, or "trophic," levels, such as invertebrates and fish.[4] Typically, about a quarter or less of the energy is derived from primary production within the stream by plants and algae, with the remaining energy coming from outside the stream or from plants on the stream edge. The upper reaches of the stream thus feed the lower reaches as the continual water movement transports food downstream.

Near a stream's source, the primary sources of organic matter are leaves, needles, and twigs. Insects play an important role in the stream ecology, processing the organic matter into smaller pieces. These insects are called shredders, grazers, collectors, and scrapers — descriptions which relate to their feeding habits. Insects dominate the invertebrate community in many streams and are responsible for most of the consumption and breakdown of the terrestrial organic matter deposited into the stream. This organic matter is rapidly colonized by bacteria. Shredders, such as cranefly and caddisfly larvae and stonefly nymph, which feed on the partially decomposed organic matter, derive nutritional benefit by ingesting these attached bacteria. The shredders break up the large pieces into small particles. The material is further broken down by collectors (e.g., blackfly and midge larvae) which filter the water or gather organic matter. Grazers forage in the sediment for food. Finally, scrapers (e.g., caddisfly larvae) remove the bottom algae from exposed surfaces.

Many of these insects are adapted to their environment by having streamlined, flattened bodies (e.g., mayfly, stonefly, and dragonfly nymphs). They often inhabit the interstitial spaces (small spaces between the sediments) of the streambed. Other insects live on the surface of the substrate. Blackfly nymphs, a common inhabitant of fast-flowing waters, attach to stones using suckers. Snails and worms are also common.

As the stream widens, more light penetrates, and the amount of primary production increases. Bottom-dwelling algae and submerged plants constitute the main primary producers of organic material in streams. Algae form mats or coverings over rocks in rapidly moving waters or can attach to the stream bottom.

Predators, both invertebrate animals, such as crustacea, and vertebrates, such as fish and amphibians, feed on the shredders, grazers, collectors, scrapers, and smaller predators. Fish increase in abundance as a stream grows in size, forming an important part of the stream's predatory community.

Fish found in streams are dependent on water depth and water flow rate.[5] They tend to have a rather specialized habitat requirement, although North American species such as carp, bluegill, smallmouth bass, and some darters and shiners are found in a variety of habitats. The presence of cover is often a factor in habitat selection, as fish will move from a preferred habitat without cover to avoid predation. Many juvenile fish, such as carp, minnows, suckers, and smallmouth bass, as well as juvenile and adult shiners, stonerollers, bluegill, darters, and sunfish are found in shallow pools (less than two feet deep) and in the slower-flowing areas.

In the faster areas of the stream, fish, such as adult darters, redhorse, stonerollers, and hog suckers, exist using the rubble substrate, boulders, and submergent vegetation as spawning habitat and cover from the current. In deeper waters (two to four feet deep), trout, adult hog suckers, bullhead catfish, bass, crappie, and stonerollers are all found. Generally, fish numbers are greatest in the slower-moving, shallower regions of streams. Shallow pools and riffles are the more common reproductive and rearing areas for fish. These are also the areas most likely to be affected by mining activity. Adult game fish prefer to inhabit the deeper, slower moving pools that are less susceptible to variations in stream flow, but these locations are not the preferred habitat of their prey — smaller fish.

At the stream edge, the type of community that develops depends on the substrate available, the slope and stability of the edge, and the constancy of the discharge. Plants consist of emergent vegetation as well as floating and submerged plants. Many insects, larvae, plankton, and fish inhabit this near shore area. In the quieter zones, more turbid, finer grained sediments are the home of burrowing mayfly and alderfly larvae, dragonfly nymphs, and annelid worms.

The stream community also includes mammals that live in close association with the direct stream community.[6] They include predators, such as otter and mink that feed mostly on fish, and herbivores (plant-eaters), such as muskrat and beaver.

PHOTO: HAGLER BAILLY SERVICES INC.

Mining pollution from snowmelt enters the uncontaminated Panther Creek, downstream of the Blackbird Mine in Idaho. This discharge was high in copper, cobalt, and suspended sediment which threatens the stream's ability to support aquatic life (see also Chapter 7, Photo).

Since a healthy stream community is often dominated by organisms that live and feed on the organic matter derived from the stream edge, the removal of vegetation, dumping of mine tailings, and other mine-related activities can disrupt the stream ecology significantly. Mining activities can directly reduce the food source for the insects and, as a consequence, the food source for fish. The insects play a vital role: without their presence, there is not sufficient food to support the larger fish. Thus the discussion of mining's impacts in the following sections focuses both on fish and insects. Insects are not only an important food source, they also have been shown to be good "indicators" of the water quality of the stream. The insects are typically more tolerant of pollution than the fish and, therefore, are often the predominant organisms remaining in mine-impacted streams. As a result, they have been the topic of a large fraction of the scientific study of polluted streams.

The Impact of Acidity on Stream Life

In many acid mine drainage streams, the concentration of acid (pH) is often so high that the streams are essentially devoid of life. Below a pH level of 4, most plants, animals, and fish are unable to survive[7] (see Chapter 1, Science Notes). There are many acid drainage streams that have acid concentrations substantially higher than this — in the range of pH 2 to 3 (i.e., ten to one hundred times more acidic than the concentration lethal to most aquatic life). However, even when concentrations of acid are non-lethal, they do have a significant impact on the stream ecology. Below a pH level of 5 (the acidity of vinegar is pH 3) most plants are severely impaired, and cattails, which are the most acid-tolerant aquatic plants, tend to predominate. Bottom-dwelling algae are similarly affected; species diversity decreases below a pH level of 5.

The tolerance of small animals for acidity is similar to that of plants. Bottom-dwelling organisms mostly disappear at pH levels between 4.5 and 5, except for the most tolerant species.[8] Amphibians are also sensitive to acidic waters. This is especially important as nearly all frog species and half of all salamander species lay their eggs in water. At pH 3.5 to 4, about half of frog and salamander embryos die.[9]

Fish are as vulnerable as other organisms to acidic water. In acid-contaminated waters, fish populations often are reduced in size and show signs that point to the eventual extinction of the local population. Acid impacts on fish are the greatest on eggs and small fry. Some studies have suggested that low pH levels impair fertilization of eggs. Fish that bury eggs are also vulnerable to low pH in the sediment interstitial waters (water that fills the pore space within the stream's sediments). In some instances, a decrease in the population of some species might lead to an increase in the size of the more acid tolerant fish species, a result of the lack of competition for food resources. Like amphibians, fish species do not appear to develop a tolerance to acidity over time.

The susceptibility of a number of common fish that might be affected by acid mine drainage in streams or lakes is shown in Table 1. The lowest recorded pH refers to the lowest pH where this species of fish has been found, while the pH at which noticeable effects (i.e., short-term visual effects) were found is also listed. Between these two values exists a range where fish will survive but will be small, deformed, and not likely to reproduce. As can be seen, a small drop in pH from 4.5 to 4.2 for perch (a twofold increase in acid concentration) would result in the eventual disappearance of the fish. Trout are more susceptible than most fish, showing effects at a pH of 6.

Acid mine drainage streams often have a pH well below 4 and often have no fish at all. The impact of acidity is immediate, as exposure to a pH of 4 for two or three days would have a significant effect on fish population, with fish deaths occurring.

Such events as heavy rains can flush large amounts of acidified mine waste into streams, causing massive fish kills. The high levels of metals made soluble by the acid water are a probable contributing factor in such kills.[10] In fact, a 1989 fish kill of 5,000 fish in the Clark Fork River in Montana, a river seriously impacted by decades of mining waste pollution (see Chapter 3, Toxic Metals), was attributed to the acidic runoff from the mine tailings associated with a summer storm. Heavy drainage from spring snowmelt and rains can generate highly acidic waters, and metals — recently deposited to the sediments during the low flow conditions of the previous summer — are readily remobilized and made soluble by the acidic spring waters.

The lowest recorded water pH where a variety of freshwater fish were found to be living, and the estimated pH at which the fish begin to experience non-lethal effects of the water's acidity.

Table 1
Survivability of Fish in Acid Water

Species	*Lowest pH Recorded*	*pH of First Recorded Effect*
Yellow Perch	4.2	4.5
Pumpkinseed	*	4.2
White Sucker	4.2	5.0
Pike	4.2	5.0
Bluegill	*	4.3
Bass	4.3	5.0
Trout	4.3	6.0
Darter	4.8	5.8
Minnow	5.7	6.0
Golden Shiner	4.7	5.2

* *These fish were not found in areas where pH fell below that of the first recorded effect.*

The acidity measure of pH is actually a measure of hydrogen ion concentration, and the effect of acidity on fish results from the interference of the hydrogen ion in water with the function of the gills. Fish gills regulate both oxygen uptake and the proper exchange of ions (such as calcium, sodium, and chloride) necessary for healthy freshwater fish.[11] The primary physiological mechanism behind fish death in acid water is a failure in body salt regulation. Acidity also affects fish uptake of oxygen, as the blood pH is lowered in response to lower pH waters. This results in a decrease in the amount of oxygen that can be stored in fish hemoglobin, and a relatively small change in blood pH can result in a large change in oxygen uptake. In short, acid water causes fish to suffer "chemically-induced suffocation."

The impact of acidic mine water on the life of streams in the vicinity of mines has been, and is, substantial. Clearly, the acidity of mine drainage will impact stream water quality and aquatic life severely, especially fish populations. Within a few miles of the source, where the pH of the water is often less than 4, no fish or other aquatic organisms are likely to be found. Further downstream, where increasing pH occurs, there will be a gradient in species composition that reflects the acidity of the water; at higher acidity levels, more acid-tolerant plants, algae, and animals will dominate, and fish populations will be impacted.

Impacts of Metals on Stream Life

Metals — such as iron, manganese, copper, zinc, cadmium, lead, aluminum, arsenic, and mercury — are typically found at elevated concentrations in streams contaminated by mining operations. Concentrations of dissolved metals in mining-contaminated streams can exceed 100 parts per billion (ppb) in some instances, concentrations that are more than one hundred times those found in uncontaminated streams (see Chapter 3, Science Notes). Metals in mine-polluted streams are often elevated at levels that approach or exceed the maximum suggested concentration for fish and other aquatic life[12] (see Chapter 3, Table). In many cases, these concentrations interfere with the growth, behavior, and reproductive success of aquatic life. However, since most streams contaminated by mining contain a suite of different metals, it is often difficult to separate the effects of an individual metal since the metals' modes of action are similar.

High metal concentrations in mining-contaminated streams result from the combination of low pH (high acidity, which dissolves heavy metals) and the direct release of the metals. These heavy metals are often associated with the sulfide mineral ores mined for precious metals. Metals released by mining activities can have long-term, extensive impacts on aquatic life. When dissolved in acidic runoff from mines, they are transported by streams and rivers. These metals gradually precipitate out (settle to the bottom) as the water becomes diluted by non-acidic water further downstream from the source of pollution. Once settled on stream bottoms, these metals can be remobilized either by a change in water chemistry or by an increase in the stream flow after a storm. Over time, these metals can migrate hundreds of miles away from their original source. The long-distance effects may not be lethal, but exposed animal species may be deformed, show impaired behavior, or they may reproduce less successfully. Therefore, the quality of stream life may suffer over an extended period of time.

Organisms are more susceptible to the impacts of metals when the metals are dissolved, because metals in their dissolved state — common in acidic mine effluent — are typically more bioavailable to organisms than they are in their non-dissolved state. Bioavailability is the measure of how much of the contaminant can be taken up by organisms under the existing set of chemical conditions.

An organism's extent of exposure to metals is associated closely with its role in the stream community (its location on the food chain and its feeding habits). For example, aquatic plants and algae, the base of the food chain, can take up metals directly from the water or from the sediment. Benthic (bottom-dwelling) algae are likely to be exposed to metals in the interstitial waters of the sediments. These concentrations at the stream bottom typically are much higher than those found in the waters above, and so it is possible that these organisms will have higher concentrations of metals than would free-floating phytoplankton. Plants with roots in the stream likewise will accumulate higher metal concentrations than plants on the stream shore. In wetlands areas, the slow flow results in increased sedimentation and in high concentrations of metals in sediments and in the wetland plants.

PHOTO: CHARLES RAY

Shown here, waste piles at the Stibnite Mine in Idaho are located immediately adjacent to the East Fork of the South Fork of the Salmon River. Runoff from mine waste piles can flow directly into this stream.

The link between metal accumulation, diet, and location on the food chain applies to stream animals as well. For example, burrowing and benthic dwellers that inhabit contaminated stream sediments are exposed to higher metal concentrations than are those organisms that feed on the surface or those that feed on plants. Higher concentrations of metals are generally found in animals that feed on detritus (dead, decaying matter often found in sediments) compared to those in animals that feed on living plant material. The chemistry of bottom sediments (such as pH) also appears to dictate how much metal is bioavailable to benthic organisms. Research suggests that as sediment pH decreases, the fraction of metals in sediments bioavailable to benthic dwellers increases.

Predators, which occupy the top of the food chain, are exposed to the metals accumulated in their prey. Again, the food source is important: those predators feeding on detritivores (animals that feed on detritus) should have higher concentrations of the metals than would predators feeding on herbivores. An increasing body of evidence suggests many metals typically found in streams as a result of mining activities, while toxic, often are not transferred efficiently from live food to predators which occupy the top of the food chain. In contrast, studies have shown that species lower in the food chain — such as benthic invertebrates feeding on plants — can take up metals more effectively than predators higher up in the food chain. Therefore, it is probable that predators such as trout and other insect-eating fish will not be exposed to concentrations of heavy metals in their food as high as those

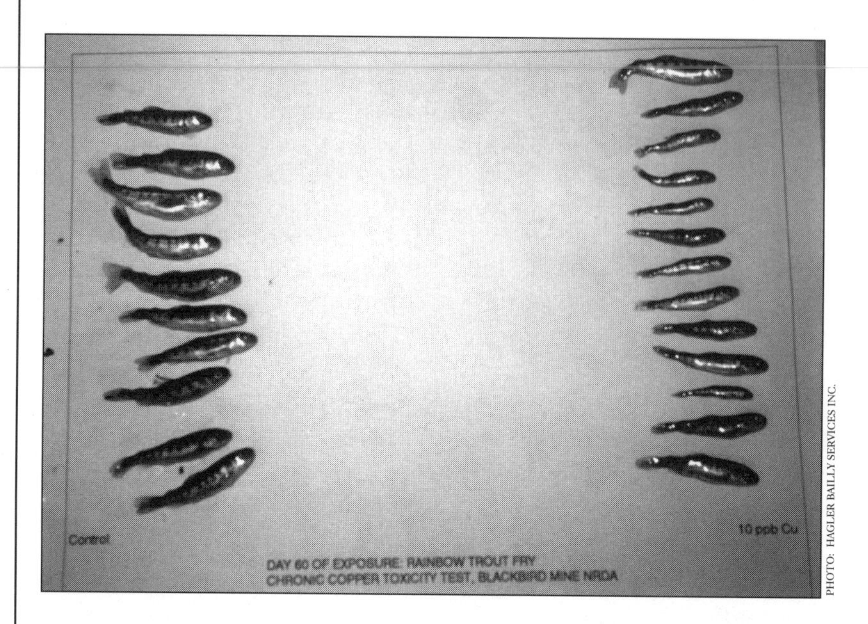

PHOTO: HAGLER BAILLY SERVICES INC.

Testing the effects of a 10 ppb copper concentration (approximately twice the ambient water quality criterion) on the growth of fish. The rainbow trout (right) show reduced growth compared to a fish from the unexposed control group (left).

associated with organisms lower in the food chain and fish that feed on sediments. This is not the case, however, for mercury (see Impacts of Arsenic and Mercury in this chapter) which is the only metal that tends to biomagnify, or become more concentrated from the base of the food chain all the way to the top.

Food, however, is not the only source of metals uptake by predators: fish also can take up metals directly from the water. Although food chain accumulation is the most likely route of uptake in most instances, concentrations of metals in the stream waters are often elevated and high enough in some instances to directly impact the organisms.[13] This can happen, for example, after spring or summer storms, when acidic water flushes large amounts of metals into streams. The result can be significant fish kills from direct exposure to metals.

In sum, heavy metal toxicity is likely to have its greatest impact on either the insects and benthic invertebrates that feed on algae or plants or those that consume sediment directly (detritivores). Fish are likely to be less impacted. However, insects and benthic invertebrates constitute the main food source of fish. As a result, the impact of heavy metals on fish will be both direct and indirect. In addition to direct contamination from food and water, the fish population may suffer indirectly because its food supply will be reduced.

Impacts of Zinc, Cadmium, Copper, Lead, and Aluminum

Concentrations of zinc, cadmium, copper, lead, and aluminum in acid mine drainage streams, in the range of 10 to 500 ppb or more, exceed the concentrations known to cause toxicity to organisms in laboratory exposure studies. For example, ambient cadmium water concentrations of more than 10 parts per billion (ppb) are associated with mortality, reduced growth, and inhibited reproduction in many aquatic species. Aquatic insects, for example, died at exposures of 3 ppb or less while zooplankton (floating, microscopic animals) died at 3 to 10 ppb. Juvenile fish and larvae died when exposed to 0.5 to 2 ppb while somewhat higher doses were fatal to adult fish (5 to 10 ppb). Sublethal effects (impairment but not death) were apparent at exposure to cadmium in the range of 0.5 to 5 ppb.[14] Similar effects were found with exposure to other heavy metals such as zinc and copper, and excessive dissolved concentrations resulted in long-term damage to gill tissue. In fact, the recommended maximum concentrations for fish survival are about 30 ppb for zinc and 5 ppb for copper.[15]

The toxicity of aluminum appears to be linked to higher levels of acidity (low pH) in streams. The toxicity of aluminum to fish increases with decreasing pH at higher pH ranges — with the maximum impact being found at a pH of around 5. At a lower pH level, toxicity is lessened to some extent but it is still important. The toxicity of aluminum results in the loss of ions from the blood working in a similar fashion to the influence of pH. Above a pH of 6, there is no aluminum effect. As different forms of aluminum in the water cross the gills at different rates, the pH effect is related to the changes in the form of the aluminum in the water at different pH's. The toxic concentration range is around 0.1 to 0.2 ppm.

Overall, the impact of metals on aquatic populations is often extensive. For example, approximately 120 miles of the Upper Clark Fork River downstream from Butte and Anaconda, Montana, have been severely contaminated by an estimated 100 million tons of mining and smelter waste dumped directly into the river and its tributaries. This area, now part of a complex of Superfund sites, has been intensively studied for mining's impacts on stream life (see Chapter 7, Effects on Ecosystems). Historically, the river was a major corridor and spawning ground for trout. While the river still harbors important fisheries, fish populations in the river's upper portions (which are contaminated by mining wastes) are below carrying capacity. Trout density in the upper part of the river near the mine was about 300 trout per mile, compared to populations of 1,500 to 2,500 per mile in uncontaminated rivers in the region. Brown trout, which are known to be more tolerant to contaminants, are dominant in the more contaminated reaches of the river. Sucker and squawfish, which are predators of juvenile trout, are also prevalent, because they have taken advantage of the niche left by the absence of the more sensitive trout species.[16]

Elevated concentrations of metals are widespread in the sediments and soils around the Upper Clark Fork area: concentrations of arsenic range from the 8 ppm background value (uncontaminated level) to 400 ppm. Elevated cadmium levels range from the background level of 0.4 ppm to 41 ppm. Copper's level ranges from the background level of 33 ppm to 7,820 ppm. Lead ranges from 26 ppm (background) to 680 ppm. Overall, 50 to 90 percent of cadmium, copper, zinc, and lead and 20 to 80 percent of the arsenic in the Clark Fork sediments are in a readily leachable phase that is considered to be bioavailable to organisms.

Research results from the Clark Fork River indicate that metal concentrations in aquatic organisms are related to feeding habit.[17] Generally, metals were highest in detritivores and lowest in predators. This result concurs with the earlier observation that higher metal concentrations in detritivores likely reflect the fact that accumulation is higher from the highly-contaminated sediment directly into the detritivore.

In the 60 kilometers of the Clark Fork River immediately below the mining sites, there were lower densities of insects, compared with densities farther away from the sites. Some types of mayflies and stoneflies were absent. These observations coincide with the measurement of higher metals concentrations in the mayflies, which are typically detritivores. Oligochaete worms and midges (a gnat-like fly) are the most pollution-tolerant, and they predominate in the more contaminated areas. These organisms accounted for 80 to 90 percent of the total population at the Clark Fork River site, compared to an unpolluted stream where oligochaetes comprised 12 percent and midges 6 percent of the invertebrate population. There was a complete absence of some of the most sensitive species from the Clark Fork River. Generally, insects closer to the mine sites, where sediments had higher metals concentrations, were more contaminated than were insects farther from the mine sites, where sediments had lower metals concentrations. In other words, metals concentrations in benthic invertebrates were a function of the metals concentrations (copper, cadmium, lead, and zinc) in the river's sediments.

Fish in the Upper Clark Fork River, like fish tested in other mine-impacted streams, have high concentrations of metals in their tissues. The nationwide average concentration for cadmium is 0.7 ppm whereas fish in the Upper Clark Fork contain up to 7 ppm, ten times the average concentration. In toxicity tests performed using sediments taken from the river, fish appeared to be relatively less sensitive to the contaminated sediments than did benthic invertebrates. In one such test, rainbow trout and zooplankton were found to be less sensitive to stream sediment than were the bottom-dwelling amphipod, a small crustacean, which exhibited reduced growth and delayed sexual maturation. Overall these toxicity study results fit with the general model of trace metal accumulation outlined earlier. They again suggest that, for contaminated sediments, benthic organisms, especially those that are detritivores, are more at risk than are those species that remain in the water column. It is noteworthy that fish eggs, which are deposited on stream sediments, seem particularly susceptible to sediment contamination.

In one study, water above metal-contaminated sediments was found not to result in significantly higher toxicity (as measured by egg survival). In contrast, the contaminated sediments themselves did indeed cause a decrease in egg survival, with egg mortality being directly correlated with the concentration of lead and copper in the sediment. This result suggests the role of metals in egg mortality.[18]

While fish appear to be somewhat less sensitive to metals than are benthic invertebrates, toxicity studies at the Clark Fork site do indicate harmful impacts. In a set of studies, rainbow trout were fed artificially contaminated shrimp that were enriched with trace metals to a level of similar extent to that of invertebrates in the Clark Fork River. At exposures of 55 ppb cadmium, 170 ppb lead, 1500 ppb zinc or 350 ppb copper in their food, no effect on survival or growth was noted in the fish, although the fish did accumulate metals in their tissues. In contrast, when adult and juvenile trout were fed invertebrates from the Clark Fork River (containing 15 to 30 ppb arsenic, 1–3 ppb cadmium, 180 to 260 ppb copper, 9 to 18 ppb lead, 320 to 700 ppb zinc), they showed scale loss, decreased survival, and metals accumulation.[19]

The above toxicity studies on fish suggest that food, not water, is the main source of the contaminants. However, detrimental effects of the water have been noted in a number of studies, especially during times of high acid runoff. For example, fish kills have been linked to spring and summer storms. The physical mixing of sediments caused by these storms can result in increased metals loading (particularly of copper) and low pH. The resuspension and washing downstream of tailings can also pose a direct threat to fish. In additional studies, mortality of juvenile fish has been observed when the fish were kept in elevated dissolved metal concentrations (2.2 ppb cadmium, 24 ppb copper, 6.4 ppb lead and 100 ppb zinc dissolved).[20]

In sum, the studies on metal toxicity in mine-impacted streams indicate that contaminated sediments rather than the water are the most important source of impact, with the insects and benthic organisms being more impacted than the fish. The greatest impacts on fish from metals contamination appear to be indirect, mostly arising from damage to the fishes' food source rather than from impacts on the fish themselves.

Impacts of Arsenic and Mercury

Elevated levels of arsenic and mercury have been found in acid mine wastes and in soils near smelters.[21] Studies indicate that both metals can be harmful to aquatic animals and plants.

Arsenic and mercury are much more volatile (readily convert to a gaseous phase) than the metals discussed earlier and are commonly transported in the atmosphere after release from smelter stacks to be deposited on land downwind. Because these metals can exist in many forms in the water and can be readily transformed between these states, they are easily solubilized and remobilized from both soil and sediment.

Background (uncontaminated) concentrations for arsenic are generally less than 10 ppb in water and less than 15 ppm in soils and sediments. The average U.S. drinking water concentration is 2.5 ppb. However, water near gold mines has been found to contain up to 1,400 ppb (140 times the background value), but more typically between 50 and 500 ppb.[22] Lake sediments downstream from one mine site had measured arsenic concentrations of 400 to 1,000 ppb.[23]

These concentrations of arsenic are comparable to those found to cause growth inhibition and mortality in aquatic organisms and plants. Studies have shown that growth inhibition in freshwater algae occurs at about 50 ppb of arsenic. For invertebrates, frog embryos were malformed and zooplankton and gastropods (slugs and snails) died after exposure to concentrations of 50

PHOTO: TEAL EYE PHOTO, MONTANA ENVIRONMENTAL INFORMATION CENTER

In 1994, tailings piles at the Vossburg Mine in Montana (shown here) had an estimated arsenic concentration of 13,000 ppm, significantly higher than the average naturally occurring background concentration of 44 ppm in the area (see also Photo Insert).

to 100 ppb. As with the other metals, there is some evidence to suggest that the accumulation of arsenic is lower in predatory insects, compared with detritivores that inhabit sediments. At one mine site, studies indicated that where concentrations of arsenic in sediments were higher, arsenic concentrations in insects were likewise higher. Insects from areas farther away, where arsenic concentrations in sediments were lower, had lower arsenic concentrations. Fish also showed behavioral impairment at similar concentrations. One study has shown that the growth of mallard ducklings was impaired by food (plant) concentrations of greater than 30 ppm, with higher concentrations leading to greater effects. Arsenic concentrations in plants near a smelter site exceeded 1,000 ppm — 30 times the concentration that reduced the growth rate of the mallard ducklings.[24]

Mercury is often associated with gold mining activity, but concentrations are generally lower than those of the other heavy metals. However, as mentioned above, mercury, unlike other metals, typically is biomagnified, or bioaccumulated such that its concentration increases as it moves up the food chain. Therefore, predatory fish often have the highest concentrations of mercury based on their body weight. This also makes mercury of particular health concern to humans.

Elevated concentrations of mercury have been found in some mining areas. In one instance, where gold mining relied on mercury amalgamation technology, high concentrations were found downstream from mining activities in the Carson River of Nevada, a portion of which is a Superfund site. Mercury amalgamation was used to extract gold and silver from the

Comstock Lode. As a result of the mining activity, the Lahontan Reservoir, downstream of the mining areas, contains about 300 tons of mercury, 1.5 tons of gold and 90 tons of silver in its sediments. Concentrations of mercury in sediments were 10 to 30 ppm, concentrations that are up to 80 times higher than those of related uncontaminated sediments. Annelid worms from the site had concentrations of 10 times background levels, and aquatic insect concentrations were 5 to 30 times higher. Fish in the Lahontan Reservoir have muscle concentrations of 2 to 4 ppm, 10 to 30 times background levels. This is higher than the Federal Drug Administration (FDA) maximum level of mercury in fish for human consumption of 1 ppm. The fish liver concentrations were up to 8 ppm. Frog and toad muscle had 2 to 3 ppm, with the liver having up to 25 ppm. Concentrations of mercury in birds were up to 100 ppm.[25] From toxicological studies, it has been shown that water concentrations as low as 0.3 ppb inhibit invertebrate reproduction and egg hatching success and impair fish physiology.[26] At fish concentrations above about 5 ppm, long-term effects and resultant death often occur.

Both arsenic and mercury can be potential heavy metal contaminants in streams and areas impacted by mining activity. The impact in streams is similar to that created by other trace metals, in that arsenic and mercury are toxic and relatively easily assimilated by organisms within the lower trophic levels of the food chain. In addition, since they are relatively volatile, they often are transported large distances, especially when emitted in airborne form from smelters, making their impact potential more widespread than that of the other trace metals.

Metal Concentrations in Terrestrial Animals and Plants

Studies suggest that metals may have harmful impacts on mammals that live in close association with the stream community and on wetland and streamside plants. Streamside mammals obtain their metals through the food chain by their consumption of stream plants and other animals. For most wetland and terrestrial species, however, there typically are no overt indications of adverse effects of metals contamination to waterfowl, amphibians, earthworms, or vegetative cover.

Impacts of Metals on Terrestrial Animals

Food sources appear to play an important role in metals accumulation in streamside mammals, such as river otter, mink, and raccoon. River otter feed almost exclusively on fish (e.g., whitefish, trout). Mink also rely on fish as their main food source, but they will eat large invertebrates and small mammals, as well. Raccoon have similar food sources but eat fish to a lesser extent. Because of these food preferences, otter and mink typically are more susceptible to aquatic pollutants, which tend to be higher overall in fish. For example, as discussed earlier, fish from the Clark Fork River have higher metals concentrations (lead, cadmium, and copper) than do fish from uncontaminated sites. In particular, zinc, cadmium, and lead levels in whitefish and trout are 5 to 10 times the national average values. A recent study demonstrated this relationship by showing that mink from the Clark Fork River had statistically significantly higher concentrations of lead, cadmium, copper, and zinc in liver and kidney, compared to mink from other nearby unpolluted rivers. This result was to be expected because of the higher concentrations of these metals in the mink's primary food (fish).[27]

Large mammals are greatly reduced in number along the Clark Fork River: otter do not exist as a viable population along the river, while mink and raccoon are significantly reduced in number. This is contrary to expectation, since there is no difference in the amount of suitable habitat between the Clark Fork River and nearby locations. It is likely, therefore, that the reductions in these mammalian populations are determined to some degree to the lack of suitable food, since trout and whitefish are reduced in number in the Clark Fork River. The lack of otter in the region compared to mink and beaver is probably explained by the different feeding habits of these mammals. Otter, which feed exclusively on fish, are more likely to be impacted by reduced fish populations than are mink and raccoon, which are more opportunistic feeders.

The high concentration of metals in these mammals probably exacerbates the impact of reduced food supply. The Clark Fork studies show that metals concentrations in mink from the Clark Fork River are all elevated, but the concentrations are not enough to cause direct toxicity (death) from any metal alone. It is possible that the combined efforts of all the different metals could cause toxicity. Dietary intake of metals by mink and otter, based on food source, suggest that the net intake is similar to that estimated, based on a food chain accumulation model, to cause toxic effects to these animals. As expected, based on the "food chain - toxic effects" hypothesis, there was no difference in the concentrations of the heavy metals in beaver (an herbivore feeding on trees which are unlikely to be directly impacted by the high concentrations of metals in the sediments) from the Clark Fork River and other sites. In summary, the overall effect of the high metals concentrations in the Clark Fork River is a reduction or elimination of the fish-eating mammals as a result of the reduced food source and the impact of the metals loading taken in with their food.

For muskrat and other herbivores, the bioavailability of the metals in soils and sediments is the factor determining the resultant metals accumulation. Muskrats, which feed on cattail roots, are much more likely to accumulate metals and be affected by heavy metal contamination than are other plant-eaters such as beavers. This difference is seen, since cattails, which are found in the higher sediment accumulation areas with higher metals concentrations, also tend to have high metal concentrations. One study showed that muskrats from a contaminated site had high concentrations of aluminum, cadmium, copper, and zinc in their livers when compared to muskrats from control (unpolluted) sites. The concentrations of metals in the muskrats were correlated with the concentrations of the metals in the sediments —

PHOTO: MINERAL POLICY CENTER

Silver Bow Creek in Montana has carried tailings down from the Butte mining district and deposited them along its banks, smothering plant life and posing a long-term threat to water quality.

which were 3 to 5 times higher in the muskrats from the polluted site. An especially high correlation was noted for cadmium. The contaminated muskrats weighed less, had lower amounts of fat, and showed a higher incidence of disease compared to the control population. Concentrations in muskrats were not enough to cause overt toxicity; however, the muskrats were in poor health as a result of the exposure and accumulation of the heavy metals.[28]

Impacts of Metals on Plants

Wetland plants, like most other plants, can take up metals from the soil and water. In most cases, plants are able to retain metals in their roots so that they do not impact the plants. However, under high exposures, these mechanisms are overwhelmed and metals move into the stems and leaves where they impact the plant's growth and survival.

Wetland and stream edge plants, such as cattails, pondweed, aquatic liverwort, and duckweed, generally do not show visible indications of adverse effects of metals contamination, such as reduced growth and survival. Of the metals, copper, cadmium, and zinc seemed potentially the most toxic in a contaminated wetland environment. Laboratory-based root growth studies have shown that metals exposure in the rooting water can inhibit growth at cadmium concentrations of 3 to 4 ppb and zinc concentrations of 1.5 to 9 ppm. Concentrations similar to those were found in the sediment interstitial waters at some of the sampling stations on the Clark Fork River.[29]

Along the Clark Fork River, significant impact to aquatic plants was found at the same locations where concentrations of metals in the sediments were the highest, where the benthic community was most impacted, and where studies with amphibians showed inhibitory effects.[30] Thus, effects on aquatic plants are likely in regions where impacts on the benthic insect population also occur.

Most plants can take up heavy metals. The cattail, a common aquatic plant, can take up metals effectively because it usually grows in marshy areas, where sediment accumulates. For example, cattails collected from a variety of sites along the Clark Fork River were found to have concentrations that ranged from 1 to 13 times background for copper, 8 to 50 times background for zinc, and 3 to 15 times background for lead. Leaves had significantly lower concentrations than roots but both were similarly elevated relative to uncontaminated sites.[31] Metals uptake has also been measured in the aquatic liverwort, another aquatic streamside plant. At one contaminated site, lead was present at a concentration of 20 ppb in the water, and the measured enrichment in the shoots of the liverwort was about a million — that is, on an equal weight basis, the concentration in the plants was a million times higher than that found in the water. In other studies, amounts as high as one half to two percent of the mass of this plant can be heavy metals after exposure to zinc, cadmium, lead, and mercury.[32]

Other studies have shown that vegetables and other plants can accumulate lead and arsenic. In these studies, plants growing in contaminated sediments near mines and smelters showed much higher concentrations of these metals (>1,000 ppm lead and >10,000 ppm arsenic) than did plants growing in less contaminated sediments farther away.[33]

While there is less information on the impact of trace metals on terrestrial animals and wetland and stream edge plants, it is still evident that contaminated acid mine drainage streams have an extensive impact. The most impacted individuals are those, such as otter and mink, that feed mostly on fish. Herbivores — with the exception of muskrats, which eat cattails that grow within the river bed — are less impacted. When coupled with the additional input of arsenic and other metals from smelter stack release, available data made it clear that mining has contributed significantly to the degradation of the immediate environment for all types and classes of animals and plants.

Erosion and Sedimentation Impacts on Streams

As Chapter 3 illustrates, mining activities past and present continue to expose large amounts of solid earthen material to the natural forces of wind and surface water drainage. As a result, much of this material is transported off the mining site into adjacent streams. Overall, mining alters soil and subsurface geological structure and disrupts surface and subsurface flow regimes. Mining activities can result in the diversion of flows into surface stream channels which are not suited for the higher stream flow. This, in turn, leads to stream bed erosion, high water turbidity, and changes in water quality.[34]

In the past in the United States (and still continuing in some countries), mine tailings were dumped directly into streams, changing the flow regime due to constriction of the original water course. High solids concentrations and the resultant water turbidity directly impact the aquatic life living in the stream. A high concentration of particles in the water causes a reduction in light penetration, which is detrimental to aquatic vegetation — both microscopic and macroscopic. The settling of the solid material can also cover and smother benthic vegetation. As a result, submerged plants and algae disappear from streams that are receiving high erosional discharges.

Large movements of sediment can result in the frequent change of the bottom characteristics at a given location — erosion followed by deposition — making these areas unsuitable for habitation by bottom-dwelling organisms. In addition, these changes lead to associated changes in the stream flow which will, in turn, affect the suitability of the location for fish and other free-swimming animals. Large deposits in the river can lead to flooding at times of high river flow. The effects described above impact fish by decreasing the amount of habitat suitable for spawning, as most fish prefer to spawn in the slower-moving, riffle areas of the stream and not in the regions where there is a high sediment load. This is because sediment movement can result in smothering and egg suffocation. Thus, the physical disturbance of stream flow and the changes in the characteristics of the streambed that result from the higher sediment load from mining can affect all levels of the stream food chain, from the animals and plants of the stream bottom to the fish.

Impacts of Cyanide

Cyanide is found in many forms in the environment, such as hydrogen cyanide, cyanide ion, and metal-cyanide complexes.[35] However, cyanide is only toxic to life in its free form, as hydrogen cyanide (HCN) and the cyanide ion (CN^-). Cyanide is absorbed rapidly if ingested internally, may be taken up through skin contact, and it is a powerful asphyxiant. At lethal doses, death results within minutes, as cyanide binds to enzymes crucial in cell respiration and other functions, affecting primarily the central nervous system (see Appendix). However, cyanide is rapidly transformed, both in the body and in the natural environment, to non-toxic metabolites, and it can evaporate from streams to the atmosphere. As cyanide does not persist as free cyanide in the natural environment, elevated concentrations are not found in waters, soil, and sediments that are removed from direct mine waste sources. Cyanide does not bioaccumulate through the food chain like heavy metals and other toxic contaminants.

Cyanide is a powerful poison to humans and animals. These warning signs were posted at a leaching heap under construction at the Crofoot/Lewis Mine, Winnemucca County, Nevada.

Algae and plants are the aquatic life most tolerant to cyanide concentrations. Most rural waters have concentrations no higher than a few ppb, while mine-related leach ponds typically have concentrations in the tens to hundreds of ppm, high enough to exclude most algal growth. Aquatic animal life, however, is very sensitive to cyanide and adverse effects are apparent at levels near 10 ppb for fish (these include reduced swimming performance and inhibited reproduction), with death occurring in the range of 30 to 100 ppb. There are numerous reports of extensive fish kills as a result of large releases of cyanide-containing mine effluent — even in the tens of thousands from one pollution incident.[36] In this incident, no cyanide was detectable in the stream several days later, and fish began to return to the stream about six months later.[37]

Many documented cyanide-related wildlife deaths have occurred at cyanide processing facilities, such as gold heap leaches and cyanide holding ponds. Recorded wildlife deaths at these facilities attest to cyanide's toxicity. Between 1980 and 1990, birds found dead near gold mining leach ponds in California, Nevada, and Arizona numbered nearly 7,000. Mammal deaths at these sites reached 520, and included a number of species, including rodents and bats as well as smaller numbers of coyote, foxes, skunks, badger, weasels, rabbits, deer, and beavers. There were over 38 reptiles and 55 amphibians counted dead as well. A more recent survey in Nevada (1986 to 1991) recorded more than 9,000 animal victims (more than 8,000 birds and about 670 mammals). Waterfowl and shorebirds accounted for more than two-thirds of the total birds killed, with about a quarter being perching birds.[38]

PHOTO: HAGLER BAILLY SERVICES INC.

This brown pelican was found dead at the end of a contaminated tailings river where it empties into the Pacific Ocean in Bahía de Ite, Peru (see Chapter 9, Ilo, Peru).

Birds, which have been significant victims of cyanide poisoning, are highly sensitive to small amounts of the toxin. Laboratory studies show that 2 to 20 mg of cyanide per kg of bird weight is lethal to a variety of birds. This is the amount present in 10 to 100 mL (2 teaspoons to half a cup) of 200 ppm cyanide solution, a concentration of cyanide comparable to that found at many of the gold processing facilities. Studies showed that the mallard, the most sensitive bird, needed to ingest only a tablespoon of a 200 ppm cyanide solution to have a lethal effect. At lower doses, birds might not die instantly, and thus counts of mortalities at mining sites likely underestimate the number of birds and mammals killed. Recent studies have shown that lower, non-lethal doses can stress migratory birds to the extent that these birds die during their flight after visiting cyanide contaminated sites.

In summary, the primary impact of cyanide is through direct contact and ingestion of cyanide-containing waters, with most cyanide-related wildlife deaths occurring at cyanide processing facilities. Cyanide contact is mostly lethal at mineral processing sites and associated waste holding ponds, as the concentration of cyanide is sufficient to kill the animals and birds that ingest the contaminated water. While animals are capable of metabolizing the cyanide and recovering from non-lethal exposures, recent research has shown that the animals are often severely weakened by the toxic insult, and death can occur due to other factors because the animal has been weakened by the exposure. Overall, contrary to the other mining contaminants, such as acids and metals, cyanide affects mostly animals and birds, with a lesser, but still significant, impact on stream life.

Mining Wastes Destroy Stream Ecology

The environmental impacts of mining are many and they affect all organisms living in streams that receive the insult of mining operations. The acidity of mine drainage can be such that many streams become unsuitable for all forms of life, except bacteria, with neither plant nor animal being able to withstand the acid waters. Heavy metals are toxic to living organisms at the levels often found in association with acid mine drainage. These heavy metals are deposited to the stream bed where they directly impact the bottom-dwelling organisms. In addition, the river remains contaminated even after the source of pollution is removed and will remain contaminated for years to decades without remediation. Mining results in the increased input of sediments and suspended material to streams. The higher turbidity and increased sedimentation affects both plants and animals, especially those that inhabit the bottom or use the stream bed for breeding purposes. Finally, the recent large-scale use of cyanide in gold and silver extraction has lead to the death of animals and birds that ingest the high concentrations of cyanide found associated with cyanide leach processes. Clearly, steps must be taken to halt environmental damage to freshwater streams from the activities of hardrock mining.

♦ ♦ ♦

Notes

1. K.W. Cummins, *Structure and function of stream ecosystems,* 24 BIOSCIENCE, p.631-641 (1974).

2. G.K. REID, ECOLOGY OF INLAND WATERS (1961).

3. Cummins, *supra* note 1.

4. REID, *supra* note 2.

5. L.P. Aadland, *Stream habitat types: Their fish assemblages and relationship to flow,* 13 NORTH AMERICAN J. FISHERIES MANAGEMENT, pp.790-806 (1993).

6. H.L. BERGMAN & M.J. SZUMSKI, FINAL REPORT ON INJURY DETERMINATION, CLARK FORK RIVER BASIN NPL SITES (1994).

7. ACID RAIN AND FISHERIES (T.A. Haines & R.E. Johnson, eds., 1982).

8. R.J. Hall, *Responses of benthic communities to episodic acid disturbances in a lake outflow stream at the experimental lakes area,* Ontario, 51 CANADIAN JOURNAL OF FISHERIES AND AQUATIC SCIENCES, pp.1877-1891 (1994).

9. ACID RAIN AND FISHERIES, *supra* note 7.

10. 13 ENVIRONMENTAL TOXICOLOGY AND CHEMISTRY, ASSORTED PAPERS FOCUSING ON THE CLARK FORK RIVER SUPERFUND SITE (1994).

11. Haines & Johnson, *supra* note 7.

12. R. EISLER, U.S. FISH AND WILDLIFE SERVICE BIOLOGICAL REPORTS FOR CADMIUM, MERCURY, CYANIDE, AND ARSENIC (1985-1991).

13. EISLER, *supra* note 12.

14. R. EISLER, U.S. FISH AND WILDLIFE SERVICE BIOLOGICAL REPORT FOR CADMIUM (1985).

15. ENVIRONMENTAL TOXICOLOGY AND CHEMISTRY, *supra* note 10.

16. Cain et al., *Aquatic insects as bioindicators of trace element contamination in cobble-bottom rivers and streams,* ENVIRONMENTAL TOXICOLOGY AND CHEMISTRY, *supra* note 10.

17. D.J. Cain et al., *Aquatic insects as bioindicators of trace element contamination in cobble-bottom rivers and streams,* 49 CANADIAN JOURNAL OF FISHERIES & AQUATIC SCIENCES, pp.2141-2154 (1992).

18. A.L. Leis & M.G. Fox, *Effect of mine tailings on the in situ survival of walleye eggs in a northern Ontario stream,* 1 ECOSCIENCE, pp.215-222 (1994).

19. ENVIRONMENTAL TOXICOLOGY AND CHEMISTRY, *supra* note 10.

20. ENVIRONMENTAL TOXICOLOGY AND CHEMISTRY, *supra* note 10.

21. HEAVY METALS IN THE ENVIRONMENT VOLUMES I & II (R.J. Allan & J. Niagu, eds., 1994).

NOTES

22. R. EISLER, U.S. FISH & WILDLIFE SERVICE BIOLOGICAL REPORT FOR ARSENIC (1988).

23. HEAVY METALS IN THE ENVIRONMENT VOLUMES I & II, *supra* note 21.

24. HEAVY METALS IN THE ENVIRONMENT VOLUMES I & II, *supra* note 21.

25. HEAVY METALS IN THE ENVIRONMENT VOLUMES I & II, *supra* note 21.

26. R. EISLER, U.S. FISH AND WILDLIFE SERVICE BIOLOGICAL REPORT FOR MERCURY (1987).

27. H.L. BERGMAN & M.J. SUMZSKI, FINAL REPORT ON INJURY DETERMINATION, CLARK FORK RIVER DETERMINATION, CLARK FORK RIVER BASIN NPL SITES (1994).

28. R.S. Halbrook et al., *Muskrat population in Virginia's Elizabeth River: Physiological condition and accumulation of environmental contaminants*, 25 ARCHIVES OF ENVIRONMENTAL CONTAMINATION & TOXICOLOGY, pp.438-445 (1993).

29. HEAVY METALS IN THE ENVIRONMENT VOLUMES I & II, *supra* note 21.

30. HEAVY METALS IN THE ENVIRONMENT VOLUMES I & II, *supra* note 21.

31. HEAVY METALS IN THE ENVIRONMENT VOLUMES I & II, *supra* note 21.

32. K. Satake et al., *Lead accumulation and location in the shoots of the aquatic liverwort in stream water at Greenside Mine, England*, 33 AQUATIC BOTANY, pp.111-122 (1989).

33. HEAVY METALS IN THE ENVIRONMENT VOLUMES I & II, *supra* note 21.

34. J. ROBBINS, LAST REFUGE (1994).

35. R. EISLER, U.S. FISH AND WILDLIFE SERVICE BIOLOGICAL REPORT FOR CYANIDE (1991).

36. EISLER, *supra* note 35.

37. G.R. LEDUC, R.C. PIERCE, AND I.R. MCCRACKEN, NATIONAL RESEARCH COUNCIL OF CANADA REPORT #NRCC 19246, (1982).

38. C.J. Henry et al., *Cyanide and migratory birds at gold mines in Nevada, USA*, 3 ECOTOXICOLOGY, pp.45-48 (1994).

♦ ♦ ♦

Impacts of Water Pollution from Mining: A Case Study

7

Hazardous Wastes from Large-scale Metal Extraction: The Clark Fork Waste Complex, MT

Johnnie Moore, Ph.D. and Samuel L. Luoma, Ph.D.

Johnnie Moore is a Professor of Geology at the University of Montana where he has been on the faculty since 1977. Much of his teaching is field/lab oriented revolving around direct student involvement in solving and examining local/regional/global environmental problems. Dr. Moore also directs the Inorganic Analyses Facility of the Murdock Environmental Biogeochemistry Laboratory. His research examines the cycling of metals and metalloids in aquatic systems — specifically the transport and fate of metals and metalloids in wetland, river and reservoir systems.

Samuel L. Luoma has worked for the U.S. Geological Survey as a project chief since 1975. His work has resulted in more than 100 publications in peer-reviewed scientific literature. Between 1985 and 1987 he was Branch Chief of Western Region Research at USGS and in 1989 he received the Distinguished Service Award from the U.S. Department of Interior. In addition to his ongoing research duties at USGS in Menlo Park, California, Dr. Luoma is also editor of the scientific journal **Marine Environmental Research.**

With *Overview* by Glenn C. Miller, Ph.D

Glenn Miller is currently a professor in the Department of Environmental and Resource Sciences at the University of Nevada, Reno. His research interests center on the fate and transport of organic and inorganic contamination in the environment. He is presently working on projects related to remediation of acid mine drainage and the impacts of precious metals mining on the Humboldt River watershed in Nevada. Dr. Miller has a Ph.D in Agricultural Chemistry.

Overview

Historic mining activity has caused major human health and environmental problems. The following article by Moore and Luoma gives a detailed and well-supported discussion of the mining impacts surrounding major mines near Butte, Montana. Mining's legacy on human health and the environment has also been felt from mines surrounding Coeur d'Alene, Idaho, as well as several other hardrock mines in many areas of North America during the past 125 years.

Human health has been impacted by toxins released into the air, from particulate and other smelter emissions, but also from contamination of surface and groundwater drinking resources. Particularly before the emergence of national environmental pollution laws and occupational health and safety laws, living or working near major hardrock mines presented health risks which would generally be considered unacceptable in today's regulatory environment. This is true not only from the view of public interest organizations, but also from the mining industry liability perspective.

The question then arises whether today's mining is still causing problems which will be considered as onerous as those of the past. The answer is not simple, since the mining industry has changed so dramatically in the last three decades. Regulation of mining has also changed and few would argue that unregulated smelter emissions such as those from the Anaconda smelter are being emitted in the same concentrations as in the past, particularly in the United States. Other poorly regulated countries, including many in the former eastern block and Asia, still have examples from present and recent past mines where it is less expensive to pay off a regulator than fix the pollution problem. In those countries, where the capability to monitor and regulate mining is lacking, more severe problems exist.

Although the type and intensity of environmental contamination from mining has changed, the ultimate costs are just as high as in the past. The reasons are two fold. First, modern mining methods have introduced new and poorly understood problems. For example, cyanide used in gold recovery is a major source of new contamination not previously encountered. Second, larger equipment and engineering techniques have allowed mining to be conducted on a much larger scale than in the past. During mining activity at the Butte site, approximately 1.3 billion tons were mined between the late 1800s and 1983. In Nevada alone, three mines larger than this one will have each excavated more rock in 20 years (1990-2010) than the total amount at the Butte site. Some of the larger modern copper mines will have extracted more than twice that amount. While rock at the Butte site may have been more acid-generating than the surface oxide ore in many gold mines, the magnitude of mining today is much greater than historic mining. Although the concentration of contaminants released is generally lower, the overall magnitude of emissions, due to the larger scale of the mines, remains a substantial problem.

PHOTO: NICHOLAS DEVORE III, BRUCE COLEMAN INC.

The Berkeley Pit, which was once the world's largest truck operated open-pit mine, encroached upon the city of Butte, Montana (see also Photo Insert).

Newer mining methods have introduced chemical problems not encountered at historic mines. The problems associated with current mines are compounded by a combination of historic contamination (i.e. acid mine drainage) and new problems, including drainage from closed heaps, leaking cyanide tailings impoundments, and the creation of large pit lakes. Each of these represents an insidious pollution problem in that they offer a very long-term source of low-level contamination. Unlike the smelter emissions of mines decades ago when people were sickened and many suffered premature death, the present pollution is more likely to remove water resources for human use, and provide a long-term source of chemical contamination to wildlife. These problems arguably will be too large to fix since most will involve millions of tons of waste rock, heaps or tailings. Examples of such problems can be seen at Hecla's Republic Mine in Washington and Independence Mining Company's Jerritt Canyon Mine in Nevada. Both have tailings impoundments that leak to surface water via underground drainage. Rinsing of the millions of tons of tailings impoundments with rain and snow will undoubtedly take decades to centuries.

Both tailings impoundments presently degrade surface water and will probably do so for a very long time. Regulators argue that these impoundments are not being constructed with the best available technology, but both were built within the past 25 years. Current mining companies are similarly able to get regulatory acceptance of proposals that are problematic, and not well understood. Too often, the burden of proof that a problem will occur is placed on the public agencies, rather than requiring the mining proponent to prove that a problem will not occur.

One of the most pressing problems in Nevada and many other western mining states is the creation of very large pit lakes, formed by open pit mining below the historic water table. The classic and most severe example is the Berkeley Pit lake, formed following cessation of open pit mining at Butte 15 years ago. In Nevada, over 35 of these lakes will remain and contain more water than all other man-made reservoirs within state borders. Most of the lakes, particularly the larger ones, will have water sufficiently degraded that it will not be suitable for human consumption, livestock watering, irrigation or wildlife use. The first large lake filling from the current mine boom, Sleeper pit lake, contains water which is approaching the quality of the Berkeley Pit lake. In the Humboldt River watershed alone, over one million acre feet of water will remain in these pit lakes, and present a contaminated legacy to humans and the environment for centuries and beyond.

Federal legislation, particularly the Clean Water Act, Safe Drinking Water Act, Clean Air Act, and mine safety laws, have undoubtedly decreased health and environment related impacts from mining, at least in the United States. We will probably not have future epidemics of disease from smelters as they existed in the past, and we will not likely have the same level of death in stream ecosystems from mining as in the past. What we can expect from the larger mining operations is very large amounts of low-level contamination over a very long period of time. This will primarily impact the health of wildlife and result in removal of drinking water sources, but it will also reduce water availability for other uses. Just as in the past, mining will continue to cause long-term environmental problems that will place a burden on future generations for remediation and for loss of public resources which could have been used elsewhere.

Abstract

Large scale metal extraction has generated extensive deposits of hazardous waste worldwide. Mining began more than 125 years ago in the Clark Fork drainage basin, western Montana, and contributed to primary, secondary, and tertiary contamination over an area one-fifth the size of Rhode Island and along hundreds of kilometers of riparian habitat. This complex of waste deposits provides numerous examples of technically difficult problems in geochemistry, hydrology, ecology, and epidemiology associated with characterizing, understanding, and managing hazardous mine wastes.

PHOTO: PHILIP M. HOCKER/ MINERAL POLICY CENTER

The Cyprus Miami copper mine is located immediately adjacent to the community of Miami, Arizona, (lower left) whose citizens are directly affected by mining activity (see photo of Bullion Plaza School in this chapter).

Introduction

The "Superfund Act" (CERCLA) of 1980 signed into Federal law the first comprehensive authority to respond to and pay for the cost of released hazardous materials into the environment (see Chapter 8). Coping with the magnitude and the diversity of the hazardous waste problems in the United States is an immense challenge, the ultimate cost of which is unknown. Others have reviewed the managerial and political challenges of hazardous waste clean-up,[1] but the technical difficulties posed by the inherently complicated nature of some contaminated sites often are not adequately considered.

A complex of waste deposits in the Clark Fork River basin of western Montana is discussed here to illustrate the number of spatially extensive, complicated problems that can develop in association with large-scale metal extraction. We describe the historic activities in the Clark Fork complex and how modern contamination is a legacy of many of those activities. An analysis of existing understanding of the contamination is accompanied by a discussion of the processes that must be better understood for effective remediation. Finally we consider whether contamination in soils, air, groundwater, and surface water threaten human and ecological health. Our conclusions point out the difficulties in remediating large scale hazardous waste problems, and thus the importance and ultimate cost effectiveness of careful waste management and waste reduction during production.

Discovery and Development

In 1805, Meriwether Lewis and William Clark began exploration of what is now Montana. Near the Clark Fork River basin, they described a "unique landscape of primitive beauty" filled with vast resources.[2] Extraction of these resources to feed the developing new nation began several decades later, and the Clark Fork River basin has supported a variety of mineral extraction activities for more than 125 years.

Placer mining for gold in the headwaters of the Clark Fork River started in 1854. Prospectors and miners pouring into Montana depleted most of the gold-bearing gravel by 1869, but discovered silver- and gold-bearing veins at Butte. Hardrock mining of these ores climaxed in 1887, when 450 metric tons per day were processed by stamp mills. When the price of silver fell in 1892, production waned and the last of the large silver smelters closed in 1896. Copper was first located in 1864. By 1896, over 4,500 metric tons of ore per day was being smelted, and construction of one of the world's largest smelting plants had begun 40 kilometers west of mining operations, at Anaconda. By the early 1910s, the new smelter was processing 11,500 metric tons of ore per day. Depressed copper prices forced closure of that smelter in 1980. In 1955, underground mining of high-grade ores in Butte was superseded by large-scale open-pit mining. Underground operations ceased in 1976. Mining of the largest open pit stopped in 1983 but has resumed in recent years along with limited underground operations.

When the smelter at Anaconda stopped production, over one billion metric tons of ore and waste rock had been produced from the Butte district. From 1880 to 1964, 297 million metric tons of ore was removed from an unrecorded amount of total material.[3] Total ore production through 1972 was 411 million metric tons, with 715 million metric tons of material removed from the Berkeley Pit between 1955 and 1973.[4] In 1973, approximately 225,000 metric tons of rock and 43,000 metric tons of ore was produced per day from the pit alone. That level of production continued until 1983, when major production stopped, accounting for an additional 675,000 metric tons of waste rock and ore.

Touted as the "richest hill on earth," Butte produced more metals than the Leadville District in Colorado or the Comstock Lode in Nevada.[5] The mining and smelting operations that produced this vast wealth left behind massive deposits of waste covering an area one-fifth the size of Rhode Island. The Clark Fork waste complex encompasses four Superfund sites, including 35 square kilometers of tailings ponds, more than 300 cubic kilometers of soil contaminated by air pollution, over 50 square kilometers of unproductive agricultural land, and hundreds of square kilometers of contaminated river bed and riparian floodplain habitat along the largest tributary of the Columbia River.

Characteristics of Contamination

Ultimately the hazardous waste problems associated with mineral extraction are determined by the characteristics of the ore and the specific processes employed to extract metals from it. The original geological studies showed that the ore body at Butte consisted of high-grade metal sulfide veins enclosed in lower-grade altered rock.[6] The predominant copper minerals were chalcocite (Cu S), bornite (Cu FeS), chalcopyrite (CuFeS), enargite (Cu AsS), and tennantite-tetrahedrite (Cu(As,Sb)S). Other associated metal sulfides included sphalerite (ZnS), pyrite (FeS), acanthite (Ag S), galena (PbS), arsenopyrite (FeAsS), and greenockite (CdS).[7] The richest vein deposits contained up to 80 percent copper and the lowest-grade, altered rock ores, 0.2 percent copper. Ores contained up to 4 percent arsenic, with some containing as much as 18 percent. Sulfur commonly exceeded 30 percent, with pyrite the most common sulfide in the ores and a primary component (0.5 to 4 percent) of the wall rock that enclosed the ores. Greenockite is rare in Butte ores, but cadmium commonly replaces other metals in sulfides (especially in sphalerite), so it is a common contaminant in Clark Fork waste deposits. These characteristics suggest antimony, arsenic, cadmium, copper, lead, and zinc should be the significant contaminants in the Clark Fork complex. Their fate also could be affected by the abundance of sulfur, especially through its role in complicated oxidation-reduction reactions.

In this paper we characterize waste products from mineral extraction as primary, secondary, or tertiary contamination. The variety of wastes produced during mining, milling, and smelting are the sources of *primary contamination.* As these contaminants are transported away from the site by water or wind, they generate *secondary contamination* in soils, groundwater, rivers, and the atmosphere. Deposits of these byproducts can be distributed over vast areas[8] and, if remobilized, can result in *tertiary contamination*[9].

Primary Contamination

The first studies of hazardous wastes in the Clark Fork complex focused on the primary contamination spread in an ill-defined patchwork of deposits over the countryside, near the modern and historic centers of mining and smelting.[10] These primary deposits contain waste rock, mill tailings, furnace slag, or flue dust. Analyses from the Clark Fork and other mineral extraction areas indicate the different types of waste have vastly different contaminant concentrations and different compositions.

Separated from ore and dumped near the mines, waste rock is probably the least contaminated material, although few analyses have been conducted. The 300 million cubic meters of rock removed from the Berkeley Pit and tens of millions cubic meters from underground workings cover approximately 10 square kilometers of land. Waste rock disposal visibly affected the countryside as early as 1912:[11]

> To one approaching the city the general appearance is most desolate. Bare, brown slopes, burnt and forbidding, from which all vegetation was long ago driven by the fumes from the smelters, rise from an almost equally barren valley. The city lies toward the base of the slopes. Within it and dotting all the hills about rise red mine buildings, which with the great heaps of gray waste rock from the mines form the most conspicuous feature of the landscape. . . Heaps of waste are everywhere prominent, attesting by their great size the extent of the underground workings.

As the ore was separated by milling and flotation, about 98 percent of it was discarded as fine-grained tailings. When the concentrate was further refined by smelting, flue dust and slag were produced. Such residues contain 100 to 1000 times natural levels of arsenic, cadmium, copper, lead, and zinc. Site characterization is a fundamental early step in contaminant remediation,[12] but locating and identifying specific deposits of these heavily contaminated wastes has been difficult because of the lack of historic records. The largest and best understood deposits occur in tailings ponds, constructed between the early 1900s and the 1950s to restrict the movement of wastes. The ponds cover at least 35 square kilometers and hold more than 200 million cubic meters of mill and smelter tailings. Based on average concentrations of metals in the tailings, approximately 9,000 metric tons of arsenic, 200 metric tons of cadmium, 90,000 metric tons of copper, 20,000 metric tons of lead, 200 metric tons of silver, and 50,000 metric tons of zinc could be present in the ponds.

Atmospheric Dispersion of Secondary Contamination

Smelter operations resulted in widespread dispersion of secondary contamination. The oldest smelting process, "heap roasting" (burning large piles of intermixed ore and timbers), released massive amounts of sulfur dioxide and metals to the atmosphere[13]. When heap roasting was prevalent in Butte in the late 1880s, the resulting fumes were quite noxious:[14]

> . . . ore was being roasted outside in the grounds of the reduction works, the fumes rising in clouds of cobalt blue, fading into gray, as it settled over the town like a pall. . . The driver reined his horse as we entered the cloud of stifling sulphur and cautiously guided them up the hill. A policeman, with a sponge over his mouth and nose to protect him from the fumes, led us to a little hotel on Broadway, for we could not see across the street.

When smelting operations were transferred to Anaconda, contamination followed. Within months of beginning production in the new smelter in 1902, outbreaks of arsenic poisoning occurred in cattle, sheep, and horses over an area of 260 square kilometers[15]. One ranch, 20 kilometers downwind of the smelter, lost 1000 cattle, 800 sheep and 20 horses during the first year of smelter operation. To reduce the damage, a flue system was constructed to settle the solids in the smoke. Even after its construction, release of 27,000 kilograms per day of arsenic, 2,300 kilograms per day of copper, 2,200 kilograms per day of lead, 2,500 kilograms per day of zinc, and 2,000 kilograms per day of antimony from the stack were documented.[16] The contamination of soils by deposition of these air pollutants was worsened when farmers were forced to irrigate with contaminated river water during dry years.[17,18] Although the extent of contamination is not completely characterized, recent estimates based upon photo reconnaissance suggest soil contamination visibly affects vegetation cover over an area of at least 300 square kilometers.[19] Thus transport processes appear to have left a legacy of secondary contamination that affects cropland, soils, and farm animals.[20]

PHOTO: ANN MAEST

A basketball court in the shadow of one of Centromin's polymetallic smelters in La Oroya, Peru. Four kilometers downstream of the smelter, soil and water samples far exceeded U.S. human health standards for arsenic, cadmium, copper and lead (see Chapter 9, Mantaro River Watershed, Peru).

Secondary Contamination of Ground Waters

Complicated reactions of the sulfur-rich Butte ores with oxygen play an important role in determining the fate of contaminants that contact water. Facilitated by bacterial decomposition, acidic waters are produced when metal sulfates react with oxygen-rich water. Through several steps, metal ions, sulfate, and hydrogen ions are produced.[21] This process mobilizes metals and metalloids previously bound in sulfide and degrades waste rock, releasing more metals into solution.

During underground and surface mining in Butte, groundwater was pumped from the workings to eliminate flooding. When open-pit mining ended in 1983, pumping was discontinued and oxygenated water began filling underground shafts and tunnels, and the 390 meter deep Berkeley Pit. These waters soon turned acidic, with pH from 2 to 3, and now concentrations of sulfate and some metals are as much as thousands of times those found in uncontaminated water. Estimates of groundwater movement suggest that 30 million liters per day of water flows into the pit, raising the water level 22 meters per year. If mean concentrations of arsenic, cadmium, copper, and zinc are 7.1,

PHOTO: MINERAL POLICY CENTER

Groundwater at the Golden Sunlight Mine, a gold operation in Montana, is less than 100 feet from the surface.

Interview:

Andrea Fine

Worn Down and the Water's Gone

Andrea Fine

Andrea Fine is currently a freelance journalist working in Philadelphia, Pennsylvania. Ms. Fine is a former newspaper reporter and served as a correspondent for **People Magazine.**

Stan and Mickey Senechal bought a home in Whitehall, Montana, in 1982. Their new home was about three miles west of the Golden Sunlight mine, a cyanide processing gold mine which was just beginning construction at the time.

Shortly after Golden Sunlight became operational, in 1983, the mine's tailings dam ruptured, and the tailings pond leaked cyanide into the groundwater. It took the company until the next year to inform the Senechals that their well — and others nearby — contained cyanide leaked from the mine. Tests also revealed nitrogen levels to be unusually high.

Here's how Mickey Senechal explained what happened next, in his words:

We were really concerned. Our son, Adam, was 14 months old at the time; Chad wasn't quite three. Just a year before, there had been that scare with cyanide injected into Tylenol pills. Of course, we'd heard about that, and this scared the daylights out of us.

We started trying to talk with them at Golden Sunlight and work something out. We didn't get an attorney right away; we wanted to work it out without a lawsuit. But then we realized we weren't going to get anywhere.

As soon as our lawsuit hit the paper, we were accused of trying to personally shut things down at the mine. Whitehall is a small town of 1,200 people, and the mine had dished out a lot of money to the schools. Half the people in town worked for them, so this got to be a personal thing. There were nasty letters to the editor about the situation.

Golden Sunlight stepped up the well tests every month. Then they disconnected the house from the well altogether and connected the house to a water tank up the hill from us. It was a 10,000-gallon tank to serve two families. The water sat there for so long, you never knew when the rust would set in. Your toothbrush would come out orange. Or, you'd go away for the weekend, and the house had a terrible stench.

We yelled and screamed at the public water bureau to come down and test the water. They finally did and found two kinds of bacteria.

We ended up getting a couple of five-gallon containers and hauling the water from town. We had a great garden up to then. Now our own well water wasn't good for outside use and our inside well wasn't good for inside use.

When the water quality bureau tested our water, they ordered Golden Sunlight to install a reverse-osmosis system. This was a real complicated piece of gear, and it would break down a lot. They'd have to repair it. The guys would hang over this open water tank in their greasy clothes. Meanwhile, we'd continue to haul water.

The lawsuit dragged on for more than five years. We were all reluctant to settle. It wasn't till a piece aired on ABC's 20/20 program about our problem that we agreed to a settlement. Our neighbors wanted that. They were having health problems because of the situation, and they just wanted out of it. We all felt we were in a constant state of anxiety.

Ultimately, as part of a court settlement, Golden Sunlight took over the Senechal's land, and the family moved to another town in Montana. Stan Senechal added the following to the discussion:

The only thing we're guilty of is getting in the way of pollution. We didn't get near the one million dollars in the suit; I'll tell you that, now. The emotional part — nobody can repair that. It got so bad I no longer liked my job. I started doubting myself, wondering if, as an engineer, I was doing this to other people.

I'm not against mining; it just needs to be done right. But to them [the mining company], the land and water means nothing; to us, the land was our home. We lived off the water. You just can't replace that.

0.54, 5.3, and 740 mg/liter in mine-shaft waters adjacent to the pit and inflow is 30 million liters per day,[22] 210 kilograms of arsenic, 15 kilograms of cadmium, 160 kilograms of copper, and 22,000 kilograms of zinc would be transported into the pit each day. The hydrology of this system is sufficiently complex that the ultimate fate of the contaminated water is uncertain. Resumed pumping, water treatment, and metal extraction may be possible, but specific economic, engineering, and waste disposal strategies remain to be demonstrated. Otherwise, the simplest scenarios suggest that the contaminated water will ultimately flow into the adjacent Butte Valley alluvial aquifer (probably by the turn of the century), and from there into Silver Bow Creek and the Clark Fork River — compounding existing contamination problems.

Groundwater contamination in a diverse expanse of tailings ponds is affected by a mix of complicated processes, mostly governed by reduction and oxidation of sulfur. The most recently constructed ponds are full of water, pH is near neutral, and sufficient organic matter is available to establish anaerobic conditions. Sulfides produced in these sediments would be expected to immobilize cadmium, copper, lead, and zinc, but contaminants with more soluble reduced forms, such as arsenic, might be released into groundwater. Such conditions occur in a contaminated reservoir at Milltown,[23] but have not been verified in the ponds. In older ponds, organic material is limited, and small inputs of water oxidize sulfides. The pH is reduced, and thus most metals could be carried into the underlying alluvial aquifer. In a pond in Butte, the metals appear to re-precipitate where they encounter a subsurface anaerobic zone rich in organic material.[24] Analyses of groundwater below the ponds at Anaconda suggest contaminant penetration is occurring there. Contaminants are found in groundwater at depths of 10 to 25 meters, and as much as 1 kilometer down-gradient. If the oxidized zone extends through the entire thickness of these tailings, or there is not sufficient organic material available for reduction, arsenic, cadmium, copper, and zinc could infiltrate into the underlying aquifer. The processes affecting groundwater contamination are understood in only the most general sense in the Clark Fork complex, thus prediction of distribution, fate, or movement of contamination has been difficult (see Chapter 5, Hydrology and the Spread of Mining Contaminants).

Secondary and Tertiary Contamination by River Transport

Because of the long-term deposition of contaminants in the system, riverine transport of secondary and tertiary contamination may be much more extensive than previously thought. Recent studies show that metals can be transported away from the primary sources as either particulates or as solutes of secondary origin. One source of the solutes is metal sulfate in the upstream floodplain soils.[25] The sulfates form as acid waters evaporate in the summer. When mixed with water, these compounds readily dissolve, pH drops to low values within seconds, and solute metal values reach many hundreds of mg/liter.[26] Thus intense rain storms can transport large amounts of dissolved metals and acid into the river.

Contaminated particulates are widely dispersed in the river system. Fine-grained sediments in the river and its reservoirs are contaminated for more than 560 kilometers downstream from the smelter.[27,28,29,30] The contamination follows a simple exponential decline that fits both river bed and reservoir sediments through this distance. Concentrations of metals in river sediment near Anaconda (at the confluence of the headwater tributaries) are twenty to more than one-hundred times higher than those in uncontaminated tributaries. At 380 kilometers, concentrations still exceed those in the least enriched tributaries by ten times or more. If the exponential function is extrapolated downstream, it suggests that detectable enrichment of most metals would extend into Pend Oreille Lake.

Much of the particulate contamination probably originated from historic mineral extraction activities that, until the 1950s, did not efficiently trap particulates before they entered the river. Until the early 1900s, much of the particulate waste material from milling and smelting in the Clark Fork complex was sluiced onto surrounding land surfaces or directly into local streams. The two tributaries in the headwaters, Silver Bow and Warm Springs creeks, transported the bulk of these wastes away from the mines and smelters (see Chapter 6, Photo). These streams, although only 0.4 percent of the total discharge of the Clark Fork River, have supplied the majority of the metallic contaminants to the drainage. Early observers noted that discharges of contaminated particulate material kept the Clark Fork River turbid over 200 kilometers downstream[31] at least periodically into the 1950s, until completion of the last tailings ponds. The addition of huge amounts of sediment to the river system plugged stream beds causing extensive flooding[32] and deposition of contaminants on the surrounding floodplain. Vast areas of the floodplain became contaminated wastelands (slickens) first described in 1917:[33]

> A trip through the region affected by the tailings presents a very interesting picture. Before their advent the soil supported the characteristic flora of this district which is still seen outside the tailing areas. . . flourishing willows line the little streams while grasses of various kinds, the wild rose, and clover among other things grow abundantly. . . altogether a typical mountain valley. In contrast, among the tailings the willows in places stand back and dead for thousands of yards at a stretch while at others they have an unhealthy appearance. . . The soil is gradually covered by the tailing solids which impart to it a variety of colors, in some cases gray, in others yellow or bright red from ferric oxide. For miles along the streams where the water is evaporated away the ground is encrusted with masses of bright blue and green deposits. . . the blue, a basic copper sulfate, and the green, a mixture of copper and iron sulfates. . . The water in many of the rivulets is decidedly acid with sulfuric acid while the rocks in the bed of the streams are mostly changed. . . into velvety pebbles of various shades of green, the color again being due to compounds of copper. Even the bones of perished stock, instead of being bleached, are dyed a vivid green.

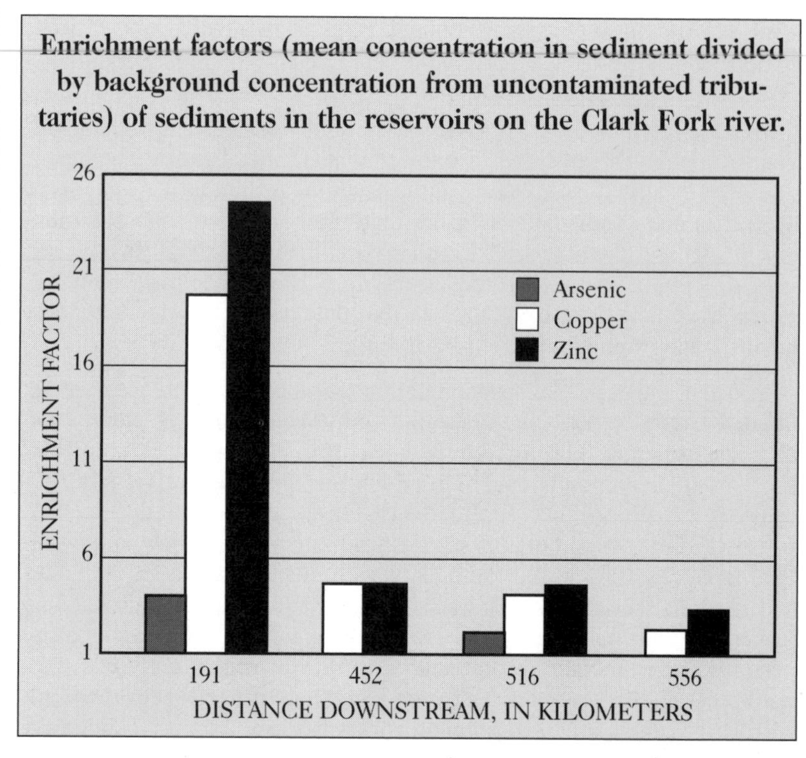

Enrichment factors (mean concentration in sediment divided by background concentration from uncontaminated tributaries) of sediments in the reservoirs on the Clark Fork river.

Not much has changed in seventy years. Slickens with malachite-colored bones can still be seen along the banks of the Clark Fork River for over 100 kilometers from its origin.

Floodplain sediments in the uppermost Clark Fork contain arsenic a few hundred times, copper a thousand times, and zinc a few thousand times background values found in uncontaminated tributaries. Highly contaminated cutbanks have been found 200 kilometers downstream.[34, 35] Johnson and Schmidt[36] suggest that approximately 1 million cubic meters of tailings reside on the floodplain between Warm Springs and Deer Lodge. However, 1.2-2.5 million cubic meters of tailings have been identified along Silver Bow Creek alone,[37] and visible patches of tailing materials also cover tens of hectares as far as 60 kilometers below Deer Lodge. These data suggest a minimum of 2 million, and likely more than 3 million cubic meters of contaminated sediments in the floodplain. This type of secondary contamination can provide a huge non-point source of metals as a river meanders through its floodplain. Continuous inputs from such a source might extend the downstream penetration of the contamination.

The distribution of metal enrichment in the floodplain is highly variable downstream.[38, 39] Processes that contribute to the variability appear to include historically variable sediment transport; spatially and temporally variable geochemical mobility from soils; highway and railroad construction that isolated patches of old floodplain or moved the river to banks unaffected by historic deposition of wastes; and perhaps historic variability in mining and smelting processes. Because of this patchiness, quantitatively evaluating the importance of bank inputs may require understanding which cutbanks specifically contribute to sediment loads or how metals are distributed among banks with differing geomorphological activity.

Dams may trap sediments in the Clark Fork, but they do not necessarily prevent downstream transport. Four dams occur on the river. The oldest was built in 1907 at Milltown, 190 kilometers downstream from the origin of the Clark Fork River. Additional reservoirs were built at 452 kilometers in 1915, at 556 kilometers in 1952, and at 516 kilometers in 1959. Elevated concentrations of at least some contaminants have been determined in all the reservoirs[40] (see Bar Graph). Furthermore, the presence of the dams does not appear to affect the downstream trend of contamination. The specific effects of the dams on the long-term fate of metal-contaminated sediments in the river clearly needs more study.

Reservoir sediments also may act as a toxicant sink, and a source of tertiary contamination of local groundwaters. A tertiary contamination problem of this type was discovered in Milltown Reservoir.[41] Although it is over 200 kilometers from the mines and smelters at Butte and Anaconda, this reservoir filled with sediments apparently released during the early stages of mining and smelting. Today it retains approximately 100 metric tons each of arsenic and lead, 13,000 metric tons of copper, and 25,000 metric tons of zinc.

Tertiary contamination of groundwater was discovered in November, 1981, when community water wells adjacent to Milltown Reservoir were found to contain arsenic levels well above the EPA drinking water standards. Oxidation-reduction processes released arsenic from the reservoir sediments contaminating the adjacent alluvial aquifer. The plume of contamination extended only a few hundred meters from the reservoir but covered an area of nearly 3 kilometers beneath and adjacent to the reservoir. When evidence showed that the health-threatening contamination originated from the adjacent reservoir sediments, the site was placed on the original Superfund National Priorities List. The aquifer was abandoned in 1981 and a new water supply for the community developed.

Effects on Ecosystems

The risk of adverse ecological effects associated with metal extraction is high because of the high concentrations in the waste of potential toxicants such as copper, zinc, cadmium, lead, and arsenic. Trout are one of the most valuable ecological resources affected by metals in the Clark Fork. Trout densities in most of the Clark Fork are only one-tenth or less of those in nearby streams of similar size and comparable habitat.[42,43] Only brown trout occur in the most contaminated reaches, in contrast to diverse assemblages of trout species found in uncontaminated waters. However, Clark Fork fish populations are not related to contaminant distributions in a simple fashion. High densities of brown trout occur in one small area in the uppermost river in the presence of some of the highest contaminant concentrations, suggesting complex processes may affect the bioavailability of the metal toxicants and trout success in different reaches of the river.

Results from caged fish studies in Panther Creek, downstream of the Blackbird Mine in Idaho, revealed that mortalities were caused by exposure to copper and cobalt from the Blackbird Mine

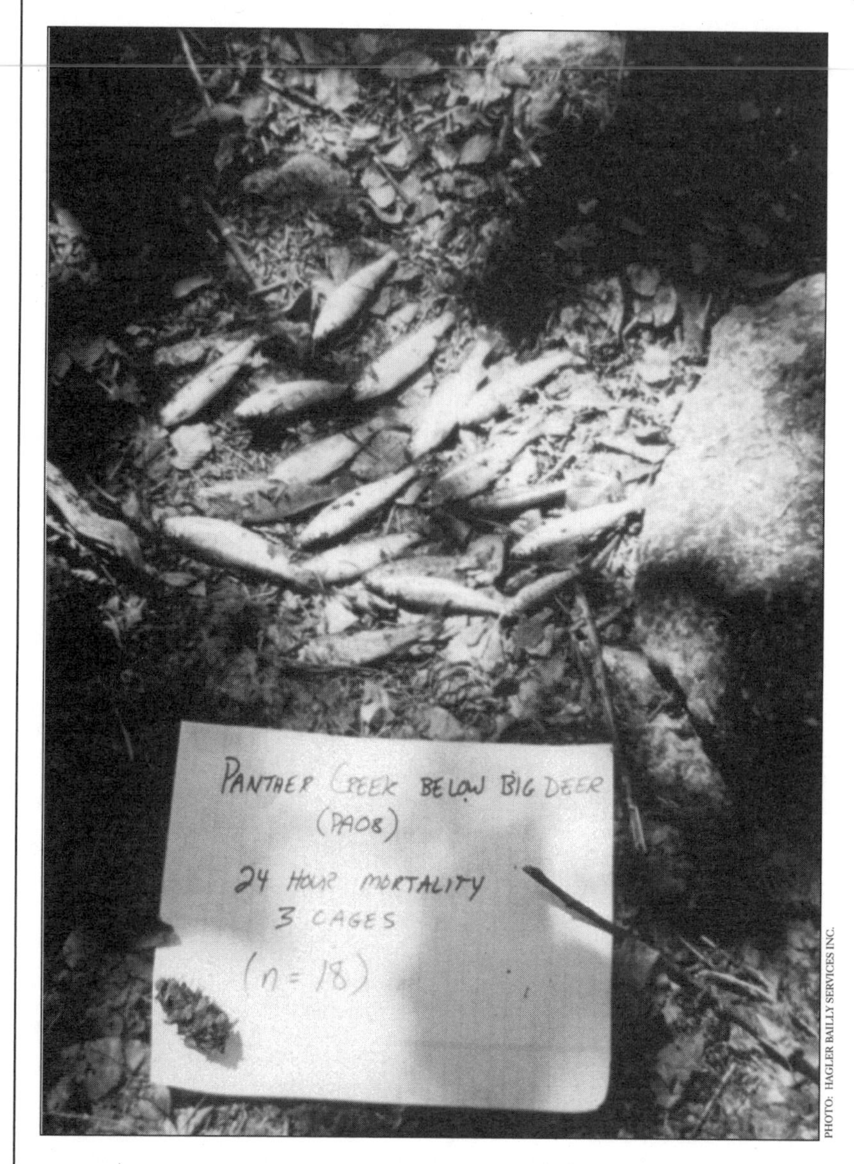

PHOTO: HAGLER BAILLY SERVICES INC.

In addition to the continuous contaminant exposures indicated by persistent sediment contamination, biota of the Clark Fork are exposed to periodic episodes of much higher contamination during some high-flow events. Acute toxicities of river water to caged trout were first demonstrated by Averett[44] during an episode in March, 1960. The toxicities coincided with "red," "high iron content," "discolored" water that occurred as far as 380 kilometers downstream from Anaconda. In more recent years, fish kills have coincided with summer storms in the upper 100 kilometers of the Clark Fork.[45] It remains unclear which water quality factors cause the fish to die so rapidly in these episodes (low pH, Fe-Al coagulates, high cadmium, copper, or zinc?). Fish also seem to return quickly in the upper river, suggesting immigration from uncontaminated tributaries might be an important process.

One initial step in assessing ecological effects of persistent contamination of the bed sediments is to determine metal concentrations in the tissues of animals that live on the river bed, many of which are crucial in the food web of fish. Recent studies show high concentrations of copper and cadmium in benthic invertebrates, especially in the Upper Clark Fork (above the Blackfoot), where fish populations are most severely reduced. In web-spinning caddisflies (*hydropsyche sp.*), at three stations between Anaconda and Deer Lodge, copper concentration was 186+36 µg/gram dry weight, cadmium 2.8+1.1 µg/gram, and lead 12.8+2.6 µg/gram. At three downstream stations, between Alberton and the Flathead Confluence, copper averaged 27+8 µg/gram dry weight, cadmium 0.7+0.3 µg/gram, and lead 3.1+1.5 µg/gram. In the least contaminated tributaries in the watershed, mean concentrations in this species were 15+1 µg/gram for copper, <0.2 µg/gram for cadmium, and 1.0 µg/gram for lead. These results demonstrate that downstream as far as 380 kilometers, contamination of sediments is passed on to biota. An extensive area of river is contaminated with biologically available metals, an observation that previous studies of effects on benthic communities and fish have not always considered.[46,47]

It should be recognized that the effects of contamination on trout and associated organisms in a river are typically expressed within the context of poorly understood environmental and ecological relationships; and conclusively demonstrating the causes of problems manifested as chronic ecological change can be difficult. Long-term, sophisticated manipulation studies have demonstrated the naivety of employing simple, singe factor analyses to explain the disappearance of large, upper trophic level species.[48] Flow, temperature, and food web characteristics, among other biological and environmental processes, interact with contaminants to determine the well-being of species. We can expect that a complete understanding of how contaminants affect trout in the Clark Fork will require careful, systematic, multi-year studies of such interacting processes. If solutions to the loss of the trout resource are possible, understanding the processes that control and affect the toxicity will be their source.

Effects on Human Health

Elevated death rates from disease are, in general, associated with historic mineral extraction areas.[49] One possible reason is that several of the contaminants typically associated with metal extraction activities are hazards to human health (see Appendix). Arsenic is a carcinogen;[50] cadmium is associated with high blood pressure and kidney disease;[51] and lead is associated with high blood pressure and behavioral anomalies in children.[52] Radon, another carcinogen, has not been studied in the complex, but is a possible contaminant because of the high uranium content in the ore body.

Several national data bases on mortality from disease include cities or counties from the Clark Fork complex, and can be employed in comparative assessments of risk of disease in the area. The national health statistics were established specifically to identify high risk localities, and to identify locali-

Ranking of Butte, Montana (relative to the 480 largest U.S. cities) for disease-caused mortality.

	1949-51	1959-61	1969-71
All Diseases	1	1	5
Heart & Kidney	1	2	31
Heart Disease	2	18	94
Other than Heart & Kidney	3	1	1

Great Falls and Billings rank between 350-450 in all categories.

Butte had the highest mortality ratio in the nation from all types of disease and from heart + kidney disease in 1949-51. Mortality ratio is the per capita mortality rate relative to that expected for the nation.

Table: Data from: M. Feinlieb et al., 1979.

ties that need more detailed study.[53] Cause and effect are difficult to determine from such statistics, although methods such as comparing rates among men and women can be employed to help separate occupational from environmental risks. Available statistical data of relevance to the Clark Fork complex include the National Cancer Institute/EPA's U.S. cancer mortality trends comparing more than 3,000 counties from 1950 to 1979[54, 55, 56] and the National Institute of Health's comparison of mortality from cardiovascular and non-cardiovascular disease in 480 U.S. cities including Butte, Great Falls, and Billings in Montana.[57]

The above data sources all indicate that the incidence of mortality from serious disease has been unusually high in the Clark Fork complex, especially in the areas where primary contamination occurs. Between 1959 and 1972, Silver Bow County was among the 100 counties in the nation with the highest mortality rates from disease for people aged 35-74.[58] The death rate in Butte from disease was the highest, or among the highest, of any city in the nation between 1949 and 1971, when adjusted for population[59] (see Table). High rates of death from heart and kidney disease in Butte contributed to the elevated mortality ratio for all diseases, but the city ranked even higher for incidence of mortality from diseases other than cardiovascular and kidney.

Comparisons of cancer rates by county also showed elevated incidence of some cancers in the Clark Fork waste complex. Counties in the area of primary contamination were among the U.S. counties with the highest rates of mortality in males and females from all types of cancer[60] and, more specifically, from trachea, bronchus, and lung cancer through 1979. Average age-

PHOTO: AIMEE BOULANGER, MINERAL POLICY CENTER

adjusted mortality rate due to the latter cancers among white males in Montana, Idaho, Wyoming, and North Dakota between 1950 and 1969 was 25+4 deaths per 100,000 people. Deaths from these diseases occurred at more than twice that rate in the counties containing primary contamination.[61] During this period, 20.5 percent of the total number of such cancer deaths in Montana occurred in these counties, among 6-7 percent of the state's population. The risks of cancer did not appear to be purely occupational. In 1970-79 death rates in women from a variety of cancers were statistically greater than the norm in the nation.[62] Overall cancer rates in Butte women were in the highest 4 percent of U.S. counties during this period.

In March of 1993, the Bullion Plaza School was determined structurally unsound, condemned, and sold to the town of Miami, Arizona. School authorities say that mining activity at the nearby Cyprus Miami copper mine (copper tailings piles, top right) contributed damage to this school.

Some statistical data suggest the incidence of lung cancer was not increasing as rapidly in the Clark Fork complex as it was in the rest of the nation in the 1970s; but in 1979 (the latest available national comparisons) risk of death from disease remained high, especially among women.

The ultimate challenge at a hazardous waste complex is to determine if the contamination in soils, air, groundwater, and surface water threaten human health. Comparisons with available national statistical data show elevated incidences of mortality from serious disease have occurred in the areas of primary contamination in the Clark Fork complex. Detailed local studies should be undertaken immediately to determine if the risk of death from disease remained unusually high into the 1980s; if such risks are environmental, or related to confounding exposures such as smoking; if elevated incidence occurs outside the areas of primary contamination; and if relationships with specific types of contaminant exposure can be established.

Strategies for Solution

Much remains to be learned about the nature and effects of hazardous wastes generated by metal extraction activities in the Clark Fork complex, but studies to date already are providing some important lessons:

(1) The long history of mineral extraction in this area has resulted in contamination of soils, groundwater, and surface water on an immense spatial scale. Reduced availability of resources (fisheries, agricultural resources) and a high incidence of disease occur coincident with contamination, especially the most severe levels.

(2) The area affected by primary contamination is large. The diversity of deposits, the scale of the deposition, poor historic documentation, and the number of analyses necessary call for a systematic approach to site characterization and careful documentation of the results of that characterization.

(3) Environmental problems may extend far beyond the boundaries of primary contamination at metal extraction sites; extensive secondary and tertiary contamination is possible. The precise extent and location of contamination of soils, agricultural crops, livestock, fish, or groundwater in the Clark Fork Basin is not yet adequately documented, but the scale is hundreds of river kilometers, hundreds of square kilometers of land, and tens of cubic kilometers of groundwater. Many studies have underestimated the extent of the problems. Perhaps because many of the secondary problems are historic, the present generation may view them as part of the "normal" terrain, failing to recognize their origin in activities as much as hundreds of kilometers away.

(4) The number of separate, significant contamination problems can easily confuse prioritization of systematic characterization and remediation processes. The problems requiring immediate attention in the Clark Fork complex are numerous: identifying if risks to human health persist, identifying sources of human exposure from among the many localities of primary contamination, defining the causes of ecological problems in the Clark Fork so the fishery of the river can be improved, defining the extent and severity of contamination of soils and agricultural products, mapping pockets of contamination in the floodplain and their susceptibility to mobilization, determining what to do about the contaminated water rapidly filling the Berkeley Pit, determining if contaminated groundwater under the older tailings ponds will spread, determining if groundwater contamination occurs under floodplains and other unstudied deposits, to name a few. Some problems are interconnected. For example, removing contaminated sediments from downstream reservoirs is futile if contaminants are continually re-supplied from contaminated floodplains. Prioritizing efforts[63] is not a trivial problem where a number of interconnected, important problems compete for limited funds. The piecemeal contracting that is common at hazardous waste sites adds to the difficulty of establishing the integrated, prioritized, systematic strategy for problem management that seems critical.

(5) Many individual problems are sufficiently complicated that solutions are not immediately obvious. In the Clark Fork many of the above problems fit this statement to some degree. The extent of the groundwater problem,

PHOTO: DAVE ZIMMERMAN

Pony Mill (top center), a gold ore processing facility, located directly upgrade from the town of Pony, Montana. Cyanide was discovered in a residential drinking water well below the mill in late 1994 (see Chapter 3, Impacts on People).

and the likely presence of sorbed phases will hinder solutions to inherently difficult groundwater cleanup efforts. Removal of primary wastes to containment areas carries unacceptable financial and ecologic costs where the area involved is 20 percent the size of Rhode Island. Restoring the river must involve dealing with hundreds of kilometers of contaminated floodplain, and manipulating a poorly understood ecological system. Defining the significance of human exposures to contamination will be limited by the area's (statistically) small population. Resolution and remediation of all the problems of the Clark Fork complex by immediate application of "proven and effective technologies"[64] seems naive. Some such "fixes" may merely relocate or even exacerbate poorly understood problems. Where mitigation of health risks (for example) appears to necessitate cleanup, but the best solutions are unclear, the efforts could be approached as full-scale, real-time experiments[65] accompanied by follow-up studies that monitor results and progressively improve approaches.

(6) Developing additional process understanding may be cost effective in solving some problems. Creative solutions to local problems and to the problems of large scale metal wastes in general will develop as understanding of these environments improves. Examples of important questions in the Clark Fork might include the following: What approaches are feasible for metal recovery from the water in the Berkeley Pit? How important is immigration in maintaining trout in the Clark Fork River, and is preservation of water quality in tributaries a critical first step in preventing further loss of the fishery? What effects do existing or proposed ponds have in providing refuges of improved water quality for trout populations? Reducing human exposures to contaminants and metal movement into the river both depend upon understanding the processes that mobilize wastes in tailings ponds, floodplains, and from surface deposits. All such suggestions require careful, rigorous scientific studies.

(7) Some contamination problems, because of their scale, intensity, or complexity, may not be amenable to remediation under foreseeable circumstances. Attaining pre-development status for the groundwater, river ecosystem, and land surfaces in the Clark Fork complex is now extremely difficult. Some problems might be improved (the fishery for example), but solutions for others, such as the extensive groundwater contamination under the tailings ponds, may involve perpetual monitoring[66] until real solutions are found. It is important to accept that some of our environmental mistakes have been so serious that they cannot be repaired. Modern society remains capable of such irreparable environmental mistakes. A principal lesson from the Clark Fork experience is that careful waste management and waste reduction during production of metal reserves is imperative. Recognition and assessment of the potential for creating highly contaminated primary wastes deposits, secondary/tertiary contamination in soil, ground and surface water, and deleterious consequences for human health and ecosystems should be a part of our mineral extraction efforts. The immense costs associated with the historic contamination of the Clark Fork Basin clearly point out the benefits of avoiding such problems in the future.

(8) The descriptors that might guide the successful approach to managing the contamination problems in the Clark Fork complex are more difficult to implement than to list. Management must be coordinated, systematic, carefully prioritized, integrated over a large area, and staffed by technically qualified individuals dedicated to the complex for the entire program. Management must be supported by studies that are multi-disciplinary, rigorously peer reviewed, systematic in their accumulation of knowledge, aware of related work, and guaranteed some continuity in support. The challenge to existing institutions is clear.

♦ ♦ ♦

Notes

1. R.A. Freeze & J.A. Cherry, *What Has Gone Wrong?*, GROUND WATER (VOLUME 27), pp.458-464 (1989).

2. W.L. Lang *in* THE LAST BEST PLACE, p.130 (W. Kittredge and A. Smith, eds., Montana Historical Society Press, 1988).

3. H.E. JOHNSON & C.L. SCHMIDT, OFFICE GOVERNOR (HELENA, MT), CLARK FORK BASIN PROJECT STATUS REPORT AND ACTION PLAN, CLARK FORK BASIN PROJECT (1988).

4. R.N. MILLER, GUIDEBOOK FOR BUTTE FIELD MEETING OF THE SOCIETY OF ECONOMIC GEOLOGISTS (18-21 August 1973).

5. W.L. Lang, *supra* note 2.

6. C.E. Meyer, P.Shea & C.C. Goddard, *Ore Deposits of Butte, Montana, in* ORE DEPOSITS OF THE U.S., 1963-1967, p.1373 (American Institute of Mining, Metallurgical, and Petroleum Engineers, 1968).

7. W.H. WEED, GEOLOGY AND ORE DEPOSITS OF BUTTE DISTRICT, MONTANA, U.S. GEOLOGICAL SURVEY, PROFESSIONAL PAPER 74, (1912).

8. T.C. Hutchinson, *Copper contamination of ecosystems caused by smelter activities, in* COPPER IN THE ENVIRONMENT, p.451 (J.O. Niagu, ed., John Wiley and Sons, New York, 1979).

9. M. Loxham, *The predictive assessment of the migration of leachate in the subsoils surrounding mine tailing and dredging spoil sites, in* ENVIRONMENTAL MANAGEMENT OF SOLID WASTE, pp.3-23 (W. Solomons and U. Forstner, eds., 1988).

10. JOHNSON & SCHMIDT, *supra* note 3.

11. WEED, *supra* note 7.

12. D.M. Mckay & J.A. Cherry, *Groundwater contamination: Pump-and-treat remediation,* ENVIRONMENTAL SCIENCE AND TECHNOLOGY (VOLUME 23), pp.630-636 (1989).

13. Hutchinson, *supra* note 8.

14. G.W. DAVIS, SKETCHES OF BUTTE: FROM VIGILANTE DAYS TO PROHIBITION (1921).

15. W.D. Harkins & R.E. Swain, *The chronic arsenical poisoning of herbivorous animals,* JOURNAL OF THE AMERICAN CHEMICAL SOCIETY (VOLUME 30), pp.928-946 (1908).

16. W.D. Harkins & R.E. Swain, *The determination of arsenic and other solid constituents of smelter smoke, with a study of the effects of high stacks and large condensing flues,* JOURNAL OF THE AMERICAN CHEMICAL SOCIETY (VOLUME 29, NO.7), pp.970-997 (1907).

17. W.G. Bateman & L.S. Wells, *Copper in the flora of a copper tailing region,* JOURNAL OF THE AMERICAN CHEMICAL SOCIETY (VOLUME 39, NO.4), pp.811-819 (1917).

NOTES

18. J.K. Haywood, *Injury to vegetation and animal life by smelter fumes,* JOURNAL OF THE AMERICAN CHEMICAL SOCIETY (VOLUME 29), pp.998-1009 (1917).

19. JOHNSON & SCHMIDT, *supra* note 3.

20. *See also* F.F. Munshower, *Cadmium accumulation in plants and animals of polluted and non-polluted grasslands,* JOURNAL OF ENVIRONMENTAL QUALITY (VOLUME 6), pp.411-413 (1977).

21. D.K. Nordstrom, *Acid Sulfate Weathering; Proceedings of a Symposium,* pp.37-48, SOIL SCIENCE SOCIETY OF AMERICA SPECIAL PUBLICATION #10, (J.A. Kittrick et al., eds., 1982).

22. JOHNSON & SCHMIDT, *supra* note 3.

23. J.N. Moore, W.H. Ficklin & C. Johns, *Partitioning of arsenic and metals in reducing sulfidic sediments,* ENVIRONMENTAL SCIENCE AND TECHNOLOGY (VOLUME 22), pp.432-437 (1988).

24. JOHNSON & SCHMIDT, *supra* note 3.

25. J.N. Moore, unpublished data.

26. D. Nimick & J.N. Moore, *Prediction of water-soluble metal concentrations in fluvially deposited tailings sediments, upper Clark Fork Valley, Montana,* APPLIED GEOCHEMISTRY (VOLUME 6, NO.6), pp.635-646 (1991).

27. C. Johns & J.N. Moore, *Copper, zinc and arsenic in bottom sediments of Clark Fork River Reservoirs- Preliminary findings, in* PROCEEDINGS OF THE CLARK FORK RIVER SYMPOSIUM, pp.74-89 (C.E. Carlson and L.L. Bahls, eds., 1985).

28. E.D. Andrews, *Longitudinal dispersion of trace metals in the Clark Fork River, Montana in* CHEMICAL QUALITIES OF WATER AND THE HYDROLOGIC CYCLE (R.C. Averett and D.M. McKnight, eds., 1987).

29. E.J. Brook & J.N. Moore, *Limitations on normalization for particle size effects in contaminated sediments,* SCIENCE OF THE TOTAL ENVIRONMENT (VOLUME 76), pp.247-251 (1988).

30. E.A. Axtmann & S.N. Luoma, *Large-scale distribution of metal contamination in the fine-grained sediments of the Clark Fork River, Montana,* APPLIED GEOCHEMISTRY (VOLUME 6), pp.75-78 (1991).

31. R.C. AVERETT, MACRO-INVERTEBRATES OF THE CLARK FORK RIVER, WATER POLLUTION CONTROL REPORT 61-1, MONTANA DEPT.HEALTH AND DEPT.FISH AND GAME, HELENA, MT, p.27 (1961).

32. O.E. MEINZER, THE WATER RESOURCES OF BUTTE, MONTANA, U.S. GEOLOGICAL SURVEY, WATER-SUPPLY PAPER 354-G, pp.79-145, (1914).

33. Bateman & Wells, *supra* note 17.

34. J.N. Moore, E.J. Brook & C. Johns, *Grain size partitioning of metals in contaminated, coarse-grained river floodplain sediment, Clark Fork River, Montana,* ENVIRONMENTAL GEOLOGY AND WATER SCIENCES (VOLUME 14), pp.107-115 (1989).

NOTES

35. Axtmann & Luoma, *supra* note 30.

36. JOHNSON & SCHMIDT, *supra* note 3.

37. ANACONDA MINERALS COMPANY, HYDROMETRICS, SUMMIT AND DEER LODGE VA., LONG-TERM ENVIRONMENTAL REHABILITATION STUDY, BUTTE-ANACONDA, VII (1983).

38. Moore et al., *supra* note 34.

39. Axtmann & Luoma, *supra* note 30.

40. Johns & Moore, *supra* note 27.

41. Moore et al., *supra* note 23.

42. G.R. Phillips, *Relationships Among Fish Populations, Metals Concentrations, and Stream Discharge in the Upper Clark Fork River, in* PROCEEDINGS OF THE CLARK FORK RIVER SYMPOSIUM, pp.57-73 (C.E. Carlson and L.L. Bahls, eds., 1985).

43. R.K. BERG, MONTANA DEPARTMENT OF FISH, WILDLIFE AND PARKS, MIDDLE CLARK FORK BASIN FISHERY MONITORING STUDY, p.39 (1986).

44. Averett, *supra* note 31.

45. Phillips, *supra* note 42, *see also* Phillips and Spoon in this publication.

46. S.P. Canton & J.W. Chadwick, *The Aquatic Invertebrates of the Upper Clark Fork River, 1972-1984, in* PROCEEDINGS OF THE CLARK FORK RIVER SYMPOSIUM, pp.46-56 (C.E. Carlson and L.L. Bahls, eds., 1985).

47. J.W. Chadwick, S.P. Canton & R.L. Dent, *Recovery of benthic invertebrate communities in Silver Bow Creek, Montana, following improved metal mine wastewater treatment,* WATER, AIR AND SOIL POLLUTION (VOLUME 22), pp.427-438 (1986).

48. D.W. Schindler, *Detecting Ecosystem Responses to Anthropogenic Stress,* CANADIAN JOURNAL OF FISHERIES AND AQUATIC SCIENCE (VOLUME 44), pp.6-21 (1987).

49. H.J. Sauer & L.E. Reed, *in* TRACE SUBSTANCES IN ENVIRONMENTAL HEALTH-XII, pp.62-71 (D.D. Hemphill, ed., 1978).

50. W.H. LEDERER & R.J. FENSTERHEIM, ARSENIC: INDUSTRIAL, BIOMEDICAL, ENVIRONMENT PERSPECTIVES (1983).

51. NATIONAL RESEARCH COUNCIL, GEOCHEMISTRY OF WATER IN RELATION TO CARDIOVASCULAR DISEASE (1979).

52. M.A. Wessel & A. Dominski, *Our Children's Daily Lead,* AMERICAN SCIENCE (VOLUME 65), pp.294-298 (1977).

53. W.B. RIGGAN, J.VAN BRUGGEN, J.F. ACQUAVELLA, J. BEAUBIER & T.J. MASON, U.S. CANCER MORTALITY RATES AND TRENDS, 1950-1979, U.S. ENVIRONMENTAL PROTECTION AGENCY PUBL.#EPA-600/1-83-0156 (1983).

54. T.J. MASON, F.W. MCKAY, R. HOOVER, W.J. BIET & J.F. FRAUMENI, 1975 ATLAS FOR CANCER MORTALITY FOR U.S. COUNTIES, 1950-1969, U.S. NATIONAL CANCER INSTITUTE, DEPT.HEW PUBL.#(NIH) 75-780 (1975).

NOTES

55. T.J. MASON & F.W. MCKAY, U.S. CANCER MORTALITY BY COUNTY, 1950-1969, DEPT.HEW PUBL.#(NIH) 74-615 (1974).

56. RIGGAN ET AL., *supra* note 53.

57. M. FEINLEIB, R. FABSITZ & A.R. SHARRELL, MORTALITY FROM CARIOVASCULAR AND NON-CARDIOVASCULAR DISEASES FOR U.S. CITIES, U.S. DEPT. HEW, DHEW, PUBL.#(NIH) 79-1453 (1979).

58. Sauer & Reed, *supra* note 49.

59. FEINLEIB ET AL., *supra* note 57.

60. MASON ET AL., *supra* note 54.

61. MASON & MCKAY, *supra* note 55.

62. RIGGAN ET AL., *supra* note 53.

63. C.C. Travis & C.B. Doty, *Superfund: A program without priorities*, ENVIRONMENTAL SCIENCE AND TECHNOLOGY (VOLUME 23), pp.1333-1334 (1989).

64. Travis & Doty, *supra* note 63.

65. Freeze & Cherry, *supra* note 1.

66. Freeze & Cherry, *supra* note 1.

♦ ♦ ♦

MINING REGULATORY PROBLEMS AND FIXES

8

L. Thomas Galloway and Karen L. Perry

L. Thomas Galloway is a leading expert in the field of regulatory enforcement of the mining industry. Mr. Galloway has served as chief litigator on numerous mining-related lawsuits of national significance, working on behalf of communities, environmental groups, and labor organizations. He also has testified before Congress on a variety of mining regulatory issues. Mr. Galloway received his law degree from the University of Virginia.

Karen L. Perry is a research assistant at the law firm of Galloway and Associates. She holds a Masters of Public Administration in Environmental Policy and Natural Resource Management from Indiana University.

Despite the devastation that hardrock mining has historically caused to water resources, and the ongoing threat of further damage, regulation of mining activities has been weak, inconsistent, and in some cases, non-existent. Although coal mining across the U.S. is governed by federal law, the hydrologic impacts of hardrock mining are not regulated in any uniform, systematic way. Instead, hardrock mining and its environmental impacts are regulated under a poor patchwork of federal and state statutes and regulations.

This chapter examines the current state of governmental regulation of the hardrock mining industry with respect to water protection. It provides a description of the patchwork system of state and federal regulatory programs, organized by statute, and illustrates the inherent problems of the current

The West Fork of Blackbird Creek, diverted into a pipe, picks up more metal contamination as it passes through this tailings dam.

PHOTO: HAGLER BAILLY SERVICES INC.

system, identifying specific gaps in the regulation of water quality impacts. Several federal statutes address, to a greater or lesser extent, the impacts of hardrock mining on water resources. Most, however, do so narrowly, resulting in a mining regulatory "system" that has *more holes than net.*

The Clean Water Act

The federal Clean Water Act (CWA) is widely considered to be one of the nation's most comprehensive and successful federal environmental statutes. In reality, the CWA as it is known today was not enacted all at once, but came about through an evolution of federal water pollution policy over a period of decades, beginning with the Water Pollution Control Act of 1948.

The Water Pollution Control Act of 1948 was the first federal statute to address what has come to be known as conventional pollutants, primarily suspended solids and oxygen demanding materials. The Act's stated goal was the "encouragement" of water pollution control.[1] It allocated funds for federal research and investigation but left regulation of water pollution to the states. Later amendments to the Act authorized states to establish water quality criteria and authorized federal funds for construction of municipal sewage treatment plants but did not address other sources of water pollution.

The Water Quality Act of 1965 required states to establish ambient water quality standards for interstate water bodies, and to implement pollution control plans designed to meet these standards. While the statute provided for federal oversight of this process, responsibility for implementation and enforcement remained with the states.

It was only with the Federal Water Pollution Control Act of 1972 (FWPCA), the heart of today's Clean Water Act, that Congress gave the federal government responsibility for setting water quality standards and for issuing discharge permits (government-issued licenses granting facilities the right to discharge wastewater off-site) and enforcing the terms of these permits. Its ambitious goals of "fishable and swimmable" waters by 1983 and the elimination of all discharges of pollutants into navigable waters by 1985 were to be achieved through the use of technology-based effluent standards developed by the U.S. Environmental Protection Agency. The FWPCA was amended five years later by the Clean Water Act of 1977, which extended some of the deadlines of the original statute and established new procedures and deadlines for determining effluent standards for toxic (as opposed to conventional) pollutants. Finally, the Water Quality Act of 1987 amended the CWA once more, further postponing the deadlines for effluent standards development. The 1987 amendments also acknowledged, for the first time, the impact of nonpoint source pollution on surface waters.

National Pollutant Discharge Elimination System

The Clean Water Act established the *National Pollutant Discharge Elimination System (NPDES)* as a method of allocating and regulating the amounts of wastes that can be discharged into the waters of the United States. This system of permitting and prescribing specific limitations on the effluent or wastewater of a facility is designed to regulate point source discharges of pollutants into surface waters such as lakes, rivers, and streams. A point source is defined by EPA regulations as a discharge from "any discernable, confined, and discrete conveyance"[2] such as a pipe, ditch, or channel. Section 301 of the CWA prohibits the point source discharge of pollutants into such waters unless authorized by an NPDES permit issued by EPA or by a state that has been delegated authority to implement the program.

The basic NPDES permit program has been of limited usefulness in controlling water pollution from hardrock mining. Mining operators are required to obtain a permit only if they allow any wastewater used in mining or processing activities to be released from the mine site by way of a point source. Many mining-related pollutant discharges do not require a permit, because they involve nonpoint source drainage — pollution that typically results from rainfall or snowmelt running off a specific site and carrying with it (mining-related) pollutants present in the area (see discussion of Federal Stormwater Regulations in this chapter).

Historically, the Clean Water Act has failed to regulate pollution of surface waters from nonpoint sources effectively. Nonpoint source pollution is caused by diffuse sources that are not regulated as point sources; it is normally associated with agricultural, urban, and construction runoff. Hardrock mining is also a major source of nonpoint source pollution, particularly in the form of acid mine drainage.

With the 1987 amendments to the CWA, the impact of nonpoint source pollution gained recognition. Section 319 of the amended statute required states to identify and assess waters impacted by nonpoint sources, determine the pollutants responsible, and begin to implement programs that address these impacts. Nonpoint sources, particularly those deriving from hardrock mining activities, however, still are largely overlooked under the CWA and continue to pollute rivers and streams throughout the United States.

The CWA and the NPDES permit program also overlook the substantial impact of hardrock mining (or any other activity) on groundwater, by failing to specifically regulate the discharge of pollutants or other impacts to groundwater. In fact, although EPA will assist states in developing their own groundwater protection programs, the agency does not require, approve, nor authorize such programs, and no comprehensive federal groundwater quality statute exists. This is a major weakness, not merely because groundwater itself is important, but also because groundwater and surface waters interact (see Chapter 5, Understanding Hydrology). In many circumstances, groundwater is a critical contributor to surface water and other river systems. Direct regulation of groundwater contamination is left to the states. While some states, such as Arizona with its Aquifer Protection program, have promulgated regulations with regard to groundwater pollution control, many have not. Even among those states that have attempted to regulate with respect to groundwater, the provisions differ greatly in approach and scope. In no case are they adequate, either as written or enforced, to protect groundwater resources from the impacts of hardrock mining.

The shortcomings of the NPDES program have allowed many hardrock mine sites to operate without permits under the Clean Water Act. In Utah, for example, the state mining regulatory agency reports that there are more than 100 large active hardrock mines, more than 300 notice mines (mines of five acres or less), and another 300 or more active mining exploration projects.[3] Despite the existence of more than 700 separate mineral activity sites, state officials currently have only *three* hardrock mines permitted under the state-implemented NPDES program.[4]

In addition to unpermitted active mines, there are many abandoned hardrock mines which are polluting water resources. Mineral Policy Center estimates that there are at least 557,650 such mines in 32 states. While many of these sites predate the CWA, responsible mining companies or landowners theoretically are still required to obtain an NPDES permit for those sites, if there is a point source discharge into the waters of the United States. As a practical matter, however, such sites do not have discharge permits. Furthermore, many sites which began mining after the establishment of the CWA were never actually permitted before abandonment. Many of these mines continue to release acidic water and heavy metals into the nation's lakes and streams. The number of abandoned mine sites is still growing, and some are capable of contaminating surface and groundwater far into the future.

Federal Stormwater Regulations

In an attempt to close one of the major gaps left by the NPDES program, EPA in 1990 published final regulations requiring permits for stormwater discharges associated with industrial manufacturing and processing activities,

PHOTO: INGA SPENCE, TOM STACK & ASSOCIATES

The contaminated water running through this tunnel deposits bright orange sediments from the Iron Mountain Mine upstream (see Chapter 1, The Threat).

including hardrock mining.[5] Under these regulations, mine operators (or landowners) are required to obtain stormwater discharge permits outlining activities to eliminate or control pollution from stormwater runoff at past or present mine sites.

Under the stormwater rules, nonpoint sources at mine sites can readily escape regulation, as they could under the basic NPDES program. Stormwater discharge permits are required only if potentially contaminated runoff drains through a defined point source. Moreover, depending on the specific nature of the pollution source, the program allows for other exemptions from these regulations.

Despite the requirement to obtain stormwater permits, very few qualified mines have applied for such permits since the regulations went into effect.

Impaired Streams

In addition to establishing NPDES permitting programs, the federal Clean Water Act requires a number of other actions on the state level to eliminate or control water pollution from industrial and municipal sources. Section 303(d) of the Act addresses the problem of lakes and streams that are *impaired* (diminished or polluted to the extent that they are not available for their designated uses, such as drinking and recreation). EPA regulations set out a process for identifying and targeting such water bodies for cleanup. The regulations require states to identify water bodies (or segments thereof) that fail to meet current water quality standards, to prioritize and target these water bodies, and to determine the Total Maximum Daily Load (TMDL), or amount per day of each relevant contaminant, that the water body can assimilate without violating water quality standards.[6]

The Total Maximum Daily Load is calculated according to a three-part formula: Point sources are assigned Waste Load Allocations (WLA); non-point sources are assigned Load Allocations (LA); and a Margin Of Safety (MOS) is used to account for the uncertainty involved in making the other allocations.[7] The Total Maximum Daily Load, then, is the sum of the WLAs for all contributing point sources, the LAs for all known nonpoint sources, and the MOS. With this formula, the total maximum allowable contamination for each impaired water body is split up and allocated among existing pollution sources.

The 303(d) listing process begins with identifying lakes and streams, or segments thereof, that do not fully meet applicable water quality standards, either numeric or descriptive, or that fully support their designated uses but are "threatened." These water bodies are classified as Water Quality Limited Segments (WQLS). After such water bodies have been identified, they are to be prioritized and targeted for TMDL development. Water bodies may be placed in one of three priority categories (high, medium, or low), based on evaluation of a number of criteria, including the magnitude of non-compliance, the size and resource value of the water body, the probability that available technology and resources can correct the problem, and the likelihood that a TMDL can be established within two years.

Where data for an impaired water body is abundant, a TMDL may be calculated and the appropriate WLA, LA, and MOS assigned in one step. Sufficient data is usually not available, however, and TMDL development typically is accomplished in phases. The results of continued monitoring and Best Management Practice (BMP) evaluations are used to revise original WLA, LA, and MOS estimates until the correct TMDL is reached. After EPA approval of the TMDL and application of necessary controls, a follow-up monitoring program is developed to ensure that water quality standards are achieved. After a period of time in which standards are met, the water body may be removed from the 303(d) list.

In theory, the TMDL process provides a means of restoring water quality where the NPDES program has proved insufficient to maintain standards. However, several recent cases point to serious problems with implementation of the TMDL provisions on the part of both EPA and the states.

In Idaho, for example, the state's 1992 WQLS list was challenged by a statewide coalition of conservation groups, who sued EPA for its failure to properly oversee the state's WQLS listing and TMDL development process.[8] After a federal district court ordered EPA to promulgate a new WQLS list, the agency found 962 threatened or degraded waters compared to the state's original list of only 36 WQLS.[9] Moreover, the court noted that the state's

1992 list was only the second list submitted by Idaho for EPA approval; the first list came in 1989, a full ten years after the statutory deadline for all states of June 26, 1979.

Although a few TMDLs now have been promulgated and approved, and Idaho is in the process of developing 29 more, the court found the state's efforts "unreasonably slow," and EPA's failure to take action as "arbitrary and capricious."[10] Consequently, in May 1995, the EPA was given one year to carry out its statutory and regulatory duty to determine, along with the state, a reasonable schedule for the development of TMDLs for all water bodies now designated as WQLS. Actual development of the more than 600 TMDLs, however, could take many years.

Unfortunately, Idaho is not unique in its failures. Although the Clean Water Act has been praised widely for its success in improving surface water quality in the United States in general, the Act has not been effective in preventing degradation of water resources from hardrock mining.

Resource Conservation and Recovery Act

The Resource Conservation and Recovery Act (RCRA) was enacted in 1976 as a set of amendments to the Solid Waste Disposal Act of 1965 (SWDA).[11] RCRA was intended to provide a "cradle-to-grave" regulatory framework to monitor and control the production, storage, transportation, and disposal of wastes deemed hazardous by EPA. The Act regulates hazardous and other solid wastes that are disposed of on land but that may ultimately create surface or groundwater contamination.

RCRA defines hazardous wastes as those solid or liquid wastes that may "cause or contribute to an increase in mortality or an increase in serious irreversible, or incapacitating reversible illness" or any wastes that may "pose a substantial threat to human health" when improperly handled.[12] While many mining wastes — including cyanide and metals-laden mine tailings — would seem to be "hazardous," the wastes produced by mining and mineral processing largely have been exempted from the hazardous waste provisions of RCRA.

RCRA requires EPA to identify and characterize hazardous wastes.[13] Acting pursuant to this mandate, the agency published a list of more than 500 substances, chemical products, and mixtures that are considered hazardous. EPA also has established four criteria — ignitability, corrosivity, reactivity, and toxicity — to determine the potential of other substances to be classified as hazardous. Normally, any waste which tests "positive" for any one of these criteria automatically falls under the hazardous waste provisions of RCRA. Certain substances, however, including most mining-related wastes, were exempted in 1980 by the Bevill Amendment to RCRA's Subtitle C (hazardous waste) provisions. This amendment to RCRA exempted from regulation several classes of low-hazard, high-volume wastes, including those produced in the extraction, beneficiation (or recovery), and processing stages of mining. The exemption was to remain in effect until the wastes could be studied further by EPA.[14] In 1986, the EPA determined, based upon its 1985 study, Report to Congress,[15] that extraction and beneficiation wastes (mostly waste rock and tailings), should continue to be exempted from regulation under Subtitle C, even though EPA found that significant quantities of these wastes exhibited hazardous characteristics. A 1991 EPA decision extended the Subtitle C program to most types of mineral processing wastes, while maintaining the exemption from Subtitle C for a subset of 20 "special", high-volume mineral processing wastes. (Mineral processing wastes are byproducts of smelting and refining, and other stages of mineral production that follow extraction and beneficiation. The "special" 20 wastes include high-volume wastes such as slag from iron, copper, and lead production as well and solid and liquid wastes generated in the production of phosphoric acid from phosphate rock). The effect of these EPA determinations has been to subject the vast majority of mining's waste stream to regulation under RCRA Subtitle D, RCRA's less-stringent standards for solid waste.

As solid wastes governed by Subtitle D of RCRA, mining wastes are treated very differently than they would be under Subtitle C.[16] While Subtitle C gives the federal government a direct role in the regulation of hazardous wastes, Subtitle D allows states to develop their own programs for solid wastes with only minimum federal guidelines. Subtitle D requirements are not independently enforceable by the federal government, and the EPA administrator has authority to bring suit on behalf of the United States only where Subtitle D violations pose *an imminent and substantial endangerment to health or the environment.*[17]

EPA has proved incapable of developing even a Subtitle D regulatory program for mining wastes. In the late 1980s, the agency undertook a process of analysis and staff-level policy discussion known as the Strawman Process, in an attempt to develop such a regulatory program. In 1990, EPA released Strawman II, a working document representing a proposed mine waste program.[18] Strawman II reflected an intent on the part of EPA to strengthen regulations pertaining to mining wastes exhibiting hazardous waste characteristics.[19] The Strawman process stalled, however, and no Subtitle D rules have ever been formally proposed or developed.

Regulation of mining waste at the state-level is based primarily on a mining reclamation program or a water pollution control program, or some combination of the two. In most states, mining wastes are not regulated under state hazardous or solid waste programs. For example, disposal of mining waste on a mine site is not regulated under Nevada's solid waste law, although a permit is required for disposal of trash associated with human occupancy of a mine site.[20] In Arizona, although mining waste technically falls within the state definition of "solid waste," it is not regulated as such in practice.[21] Instead, "material[s] produced in connection with a mining or metallurgical operation" generally are exempt from the state's solid waste plan approval requirements[22] and are not considered "toxic substances" subject to pollution prevention planning requirements.[23]

The size and quantity of mine waste piles like those shown here at the Molycorp molybdenum mine in Questa, New Mexico, present a potentially enormous pollution problem.

Comprehensive Environmental Response, Compensation, and Liability Act

The Comprehensive Environmental Response, Compensation, and Liability Act (CERCLA), better known as Superfund, was enacted in 1980 in response to growing public concern about abandoned toxic and hazardous waste sites. Whereas the intent of RCRA is primarily to prevent future contamination from ongoing hazardous waste management, CERCLA was designed to address the problems of cleaning up past pollution and preventing further pollution from inactive and abandoned sites.

CERCLA empowers the federal EPA to identify Potentially Responsible Parties (parties who may be liable for a waste site) and force them to take cleanup actions at sites the agency has targeted on its National Priorities List (NPL). As an alternative, the Act also gives EPA authority to clean up such sites itself and find or sue responsible parties for cleanup liability.

After a site is placed on the NPL, EPA conducts what is known as a Remedial Investigation (RI) of the site in order to determine the nature, extent, and sources of contamination. The next step is the Feasibility Study (FS), in which possible cleanup alternatives are identified and evaluated. After a formal public comment period on the proposed cleanup alternatives in the FS, EPA selects a final cleanup plan and formalizes it through a Record of Decision (ROD). Design and actual cleanup activities can then proceed.

Of the 1,200 contaminated sites currently on the National Priorities List, 66 are mining-related sites.[24] Many of these sites are old mines which operated before the provisions of the CWA and RCRA went into effect. Unfortunately, the mining sites undergoing and awaiting cleanup under CERCLA represent only a fraction of the many thousands of potentially hazardous inactive and abandoned mine sites estimated to exist across the country.

Hardrock mine owners and operators may be held responsible under CERCLA for the actual release or threat of release of hazardous substances — such as acid mine drainage or wastes containing heavy metals — into surface and groundwater. In addition, mine owners or operators may be held liable under CERCLA for damage to natural resources, including rivers and streams, that result from their mining activities. State or federal agencies, or other natural resource trustees, may bring damage actions against mining companies for "injury to, destruction of, or loss of natural resources, including the reasonable costs of assessing such injury, destruction, or loss." [25]

The wastes left to contaminate old hardrock mine sites sometimes can be reprocessed or remined. Where such wastes constitute a potential CERCLA liability, however, reminers may be reluctant to undertake any activity for fear of becoming liable themselves under this statute. Congress recently has considered establishing different standards for the remining of contaminated mine sites. The proposals put forward could exempt reminers from liability at CERCLA sites, not only for past damage but also for continuing damage as a result of the remining.

A major problem in applying CERCLA to hazardous mine sites is that, very often, no responsible parties can be found for such sites. In many cases, particularly where mines have been abandoned for many years or were relatively small operations in remote areas, no government agency — federal or state — has an official record of a mine's existence, much less the ability to locate former mine owners or operators.

Moreover, the Superfund program is overburdened with potentially hazardous sites. Only the largest and most severely contaminated sites make it to the NPL, while thousands upon thousands of contaminated hardrock mines continue to poison the streams of the West quietly, without any real hope for reclamation.

Statutes Regulating Activities on Federal Lands

There are several federal laws that regulate various uses, activities, and management of federal public lands.

PHOTO: MINERAL POLICY CENTER

At Bull Canyon, Monogram Mesa, the landscape has been severely scarred by mineral exploration and uranium mining.

The General Mining Law of 1872

The General Mining Law of 1872 is the chief federal statute for the regulation of hardrock mining. It contains no environmental provisions whatsoever. Now more than 100 years old, the Mining Law is a product of its time. Settlement of the West and development of the nation's resources — including land, timber, and minerals — was an important goal in the mid-nineteenth century, and federal regulation for environmental protection was essentially moot. Thus, the Mining Law was not intended to regulate the environmental impacts of hardrock mining; rather, it was enacted primarily to govern access to mineral rights and facilitate the disposal of mineral lands.

The Mining Law declared that public lands belonging to the United States (the public domain) shall be *free and open* for private mineral exploration and development; it set forth a mining claims system for allocating private property rights to hardrock minerals on such lands. Under the law, rights to locatable minerals (hardrock minerals, including metals and certain non-metallic minerals and mineral materials) on public lands can be claimed, and these claims may be registered with the federal government upon discovery of valuable mineral deposits. A valid claim gives the claimant legal title to the subsurface mineral resources and guarantees the claimant reasonable access to, and use of, the site surface for mining purposes. Furthermore, these claims can be patented — a process giving the patent holder full rights to both the minerals and the land on the surface — for as little as $2.50 to $5.00 per acre.

As recently as 1970, with the passage of the Mining and Minerals Policy Act,[26] federal policy on hardrock mining on public lands remained heavily weighted toward promoting private enterprise in the development of the nation's mineral resources. Although the Act encouraged the development of reclamation and mining waste disposal methods "so as to lessen any adverse impact of mineral extraction and processing upon the physical environment," the overriding federal policy in the statute was the promotion of "economically sound and stable domestic mining, minerals, metal and mineral reclamation activities."[27]

Because the federal mining laws are virtually silent on the issue of the environmental impacts of hardrock mining, regulation of such impacts is left primarily to the states, even though the impact is often to federal lands. In 1988, the U.S. General Accounting Office (GAO) reported that more than 424,000 acres of federal lands disturbed by mining had been left unreclaimed.[28] Active, authorized mining operations accounted for about one-third of this total acreage, while abandoned, suspended, and unauthorized operations accounted for the rest. The potential threat to water resources was significant: GAO found that almost 25,000 acres of these unreclaimed lands were in need of reclamation measures to remove or dispose of mine waste and other harmful material, while another 74,236 acres needed reclamation to control erosion, landslides, and water runoff.

The Federal Land Policy and Management Act of 1976

The Federal Land Policy and Management Act of 1976 (FLPMA) regulates all aspects of natural resources management on federal lands, including hardrock mining. The primary federal agency responsible for carrying out the provisions of this statute with regard to hardrock mining on federal lands is the U.S. Department of the Interior's Bureau of Land Management (BLM) — which has ultimate authority over all federally-owned mineral resources. Although BLM has long had the authority to regulate activities affecting public lands under its jurisdictions, regulations governing mining activities were not adopted until 1980, four years after the establishment of FLPMA.

In enacting FLPMA, Congress mandated that all activities on public lands be conducted so as to prevent *unnecessary or undue degradation* of these lands.[29] The statute requires BLM to develop Resource Management Plans and other land use planning documents for each of the agency's planning areas. With regard to hardrock mining on BLM lands, FLPMA instituted procedures for withdrawing public lands from mineral development and required mineral claims to be filed with state BLM offices.

BLM's surface management regulations require that hardrock mining and exploration activities be conducted under a Plan of Operations approved by the BLM. Operations that disturb five acres or less in a calendar year, however, are required only to file a Notice of Operation. BLM approval of this Notice is not required. In the first ten years after the regulations went into effect, BLM accepted approximately 20,800 Notices and processed 5,000 Plans.[30]

Reclamation is required for all surface-disturbing activities on BLM lands. The agency's Solid Minerals Reclamation Handbook states that reclamation standards shall be determined on a site-specific basis, using an interdisciplinary

PHOTO: PHILIP M. HOCKER/MINERAL POLICY CENTER

This moonscape is the result of several failed attempts by the Forest Service to revegetate an old mine site at the head of Fisher Creek in the Gallatin National Forest, Montana.

approach to analyze *physical, biological, climatic, and other site characteristics,*[31] and that standards *shall not conflict* with BLM Resource Management Plans. The federal regulations do not, however, set out any minimum standards for reclamation of hardrock mines. Moreover, recent BLM evaluations of hardrock mine sites on public lands have found that reclamation at such sites is often inadequate and sometimes does not occur at all.[32] This failure has been linked to inadequate inspections, lack of penalties within the regulations, and abandonment of operations with inadequate or no reclamation bonds.[33]

As part of the BLM's 1989 Surface Management Initiative, BLM offices were directed to inspect all hardrock operations at least twice a year, and producing operations using cyanide a minimum of four times per year.[34] An internal evaluation of inspection frequency, however, found that most BLM regional offices are not meeting this inspection frequency, with only 47 percent of the required inspections performed in some areas of Nevada and a mere 15 percent of the required inspections conducted in California.[35] Inadequate funding and staffing has been blamed.

BLM's record of enforcement of its surface mining regulations is also poor. Bureau staff have indicated that, because there are few serious penalties associated with a mine operator's failure to comply with BLM-issued notices of noncompliance (NONs), such notices are considered ineffective and therefore seldom used.[36]

The National Forest Management Act of 1976

The U.S. Forest Service's (USFS) regulatory program for hardrock mining operations basically parallels the BLM program, but with a few significant differences. The National Forest Management Act requires the Forest Service (an agency of the U.S. Department of Agriculture) to institute a comprehensive, interdisciplinary planning process, similar to the BLM planning process, for National Forest lands. Approximately 140 million of the National Forest system's 198 million acres are public domain lands open to mineral exploration and development.[37]

Regulations for mining on National Forest lands were late coming: they were not adopted by the Forest Service until 1974. Anyone planning to carry out mining activities on Forest Service lands must submit a Notice of Intent to the District Ranger with jurisdiction over the affected area. Unlike the BLM's mining regulations, the Forest Service regulations do not include an automatic exemption for mines disturbing less than five acres. Instead, the USFS requires a Plan of Operation for any operation where the District Ranger determines *significant disturbance* of the surface resources is likely to occur.[38]

Because BLM and USFS lands are spread across the western states, and because many of these states have their own hardrock mining regulatory programs, federal and state regulations often overlap. For this reason, both BLM and Forest Service regulations permit the federal agencies to enter into agreements for joint federal-state programs for administration and enforcement of mining regulations. These agreements are intended to prevent *unnecessary administrative delay* and avoid *duplication of administration and enforcement of laws.*[39]

The division of federal-state responsibilities is accomplished through agreements known as Memoranda of Understanding (MOUs), most of which delegate primary authority for administration and enforcement of mining regulations to the states, with BLM maintaining an oversight role. For example, under an MOU with the state of Colorado, the state's Mined Land Review Board assumes primary responsibility for the administration, review, and permitting of mining operations, as well as for conducting inspections and taking enforcement actions. Reclamation bonds are held by the state, with the provision that bonds not be released until BLM has found the reclamation to be satisfactory.[40] As of December 1994, at least nine western states had signed MOUs with BLM.[41]

National Wilderness Preservation System

Both BLM and USFS administer lands in the National Wilderness Preservation System, created by the Wilderness Act of 1964.[42] Although the law is intended to preserve "the enduring resource of wilderness" and mandates that wilderness-designated lands be left unimpaired and protected for future use and enjoyment, hardrock mining is not entirely prohibited in wilderness areas. Existing mining claims as of December 31, 1983, the date wilderness areas were withdrawn from mineral development, may continue to be mined, and the Chief of the Forest Service must allow any activity, including prospecting, for the purpose of gathering information about minerals in the wilderness areas of the National Forest system. However, mining operations in wilderness areas are more strictly regulated by both agencies than are operations in non-wilderness lands. For instance, in wilderness areas the construction of roads are limited and miners are required to conduct operations in such a manner as to protect surface resources and the wilderness character of the area.

Prospecting and mining also are prohibited in the National Parks and other areas administered by the National Park Service (NPS), although, as in wilderness areas, existing valid claims in areas subsequently brought under jurisdiction of the NPS may continue to be developed. The greatest mining-related problem in the National Parks is posed not by active mining operations, however, but by abandoned mines. The Park Service estimates that there are more than 4,000 abandoned mines in the system in 45 states; approximately 33,000 disturbed acres remain unreclaimed.[43]

State Mining and Water Programs

The absence of a comprehensive national program for regulating hardrock mining leaves the primary responsibility for this task to the states. Within each mining state, the responsibility for regulating the water quality impacts of hardrock mining usually is shared between two agencies. The first is a mining and reclamation agency that permits and oversees basic mining activities. The second is a water quality agency that assumes the primary enforcement role for implementing the CWA within the state. Depending upon the state, these agencies may or may not work together closely where their interests overlap; most often, however, there is little interaction between agencies. In addition, state water quality statutes (under the CWA or other state water laws) rarely address mining specifically or in great detail, while state mining statutes tend to overlook the long-term impacts of mining on water resources. Thus, water impacts from mining often fall through the cracks of the state regulatory structure.

State hardrock mining agencies are generally small and tend to be among the most understaffed of all environmental regulatory agencies. A Mineral Policy Center survey of eight western states in May 1994 found that the ratio of mine inspectors to active mining operations in some states may be worse than 1 inspector per 100 mine sites. As a result, most hardrock mines are inspected by state inspectors infrequently, if at all. Some states do have an inspection frequency policy; of those that do, the frequency is oftentimes less stringent than the federal policy. Montana's policy, for example, sets inspection frequency at once per year for large mines and only once every four years for notice mines of five acres or less. Most states do not even have enough basic personnel to conduct simple on-the-ground inspections, much less to conduct the sophisticated and time-consuming hydrologic analyses that a meaningful water protection program requires. Even where state laws require monitoring and reporting of water quality data, such data is often useless to the regulatory process, as states do not have sufficient or properly trained personnel to evaluate the data and take appropriate regulatory action.

Few Inspectors, Little Enforcement

The results of the MPC survey of state inspection and enforcement resources are summarized below.[44]

Arizona

- Employs 13 inspectors (from the state's Mine Inspector's Office and the Department of Environmental Quality), some with restricted duties, responsible for state inspections at 538 mines.
- Has no set schedule or policy for frequency of mine inspections.

Colorado

- Employs 15 inspectors (at the Department of Minerals and Geology) responsible for 1,944 active operations (204 hardrock or metal mines, 1,703 sand, gravel, and aggregate mines, and 37 other sites).
- Conducted only 740 inspections in 1993.
- Inspectors cited only 87 violations in 1993.

Idaho

- Employs between 3 and 6 inspectors (at the Department of Environmental Quality and the Department of State Lands) to inspect approximately 65 active hardrock mines.
- Inspectors cite on average only about two violations per year.

Montana

- Employs 20 "employees" (at the Department of State Lands) to handle permitting, administration, and other duties, in addition to inspection of 100 active mining operations and 1,000 notice mines (5 acres or less).

Nevada

- Employs 13 "employees" (at the Bureau of Mining Regulation and Reclamation) to conduct inspections, permitting, and other related activities for 225 active operations.[45]

- Inspectors cited only 6 violations in 1993.

New Mexico

- Employs 7 "employees" (at the Mining and Minerals Division) to conduct inspections, permitting, and administration for at least 185 active mine sites.[46, 47]

Utah

- Employs 4 inspectors (at the Division of Oil, Gas, and Mining) for 106 large active hardrock mines, 334 notice mines, and 326 mining exploration projects.
- Allows sites to be inspected less than once every two years.
- Inspectors cite approximately 20 violations per year.

Added to the problems of state mining and reclamation agencies are deficiencies in state water agencies. Since these water agencies generally work on many other competing interests, the percentage of funds and personnel in state water regulatory agencies that is devoted to mining-related issues tends to be small. The hands of state agencies are further tied by the close relationship the mining industry enjoys with many state governments, particularly where mining plays a significant role in the employment and economy of the state.

Weak Regulatory Standards

The states have failed for the most part to undertake the complex monitoring and evaluation of hardrock mines which is necessary to prevent environmental damage, particularly where groundwater impacts are concerned. Few, if any, states have adequate systems, technical training, and personnel to evaluate and monitor critical performance areas properly.

No state evaluates effectively and predicts accurately the potential for long-term acid mine drainage during the initial permitting process. Once sites are permitted, almost all states fail to require bonds for long-term water treatment — either adequately or at all. Furthermore, no state has the type of comprehensive enforcement mechanisms necessary to carry out these responsibilities adequately.

Finally, numerous states have enacted broad "grandfather" provisions, the effect of which is to exempt existing mines from compliance with the normally applicable standards, leaving these mines to comply with little or nothing in the way of environmental standards. In addition, many states allow for "variances" from established standards or rules under certain site-specific circumstances.

A look at the regulatory programs of several states illustrates the problems associated with state-led regulation: Arizona and New Mexico very recently enacted their first hardrock reclamation statutes, despite a long history of mining that predates statehood. Their regulatory programs are representative

The booming mineral industry in the state of Arizona poses a long-term threat to groundwater and surface water. Shown here, a copper leaching operation uses sulfuric acid.

PHOTO: PHILIP M. HOCKER/ MINERAL POLICY CENTER

of modern state regulation of hardrock mining in states where it is a growing industry. In addition, Colorado and Montana provide further examples of the problems of state mining and water programs.

Arizona. In spite of a booming minerals industry, Arizona had no mined land reclamation statute until 1994. Previously, the Aquifer Protection Permit program had been the cornerstone of the state's mine regulation "program," with the Office of the State Mining Inspector carrying out some environmental regulatory functions as an offshoot of its mine safety program.[48]

Arizona's Mined Land Reclamation Act of 1994, however, is not a significant improvement. It is rife with broad, vague regulatory language and riddled with exemptions with which mine operators may escape meaningful regulatory mandates. The statute mandates that reclamation standards be site-specific rather than standardized, directing the Director of Environmental Quality to consider the *technical and economic practicability* of proposed reclamation measures, taking into account the site-specific circumstances and the proposed post-mining land use objectives for each site.[49]

The new mining law contains a number of exemptions that hamper its effectiveness. For example, the provisions of this law do not apply to mining activities on state lands.[50] Those are regulated instead by the State Land Department. Activities at inactive mines are also exempt from the reclamation law as long as "new surface disturbances" are not created.[51] Finally, the reclamation law only sometimes applies to mining activities that create a surface disturbance of five contiguous acres or less.

Arizona's reclamation law also provides for a broad range of variances from rules that may be promulgated under it. For example, *The director may grant … a conditional order allowing [a mine] to vary from any rule adopted pursuant to this chapter or any requirement or condition of a reclamation plan ... if the ... order will not endanger public safety.*[52]

A number of mines in Arizona have experienced ongoing problems of groundwater and surface water contamination from leach heaps and tailings impoundments, a strong indication of the inability of Arizona's regulatory programs to protect water quality from the impacts of hardrock mining.

New Mexico. New Mexico passed its first hardrock mine reclamation law in 1993 and adopted rules under the New Mexico Mining Act (NMMA) in June of 1994. Under the NMMA, new mining operations — those begun after June 18, 1993 — cannot receive operating permits without approved reclamation plans and posted financial assurance (reclamation bonds). The law provides for an abbreviated permitting process for mining operations with *minimal impact on the environment.*[53] These operations are only required to submit general plans, as opposed to the more detailed and specific application documents required of other operations.

Reclamation under the NMMA is required to achieve a *self-sustaining ecosystem or post-mining land use,*[54] yet this vague requirement may be waived for an open pit or waste unit if it is *technically or economically infeasible or environmentally unsound,* provided that, *measures will be taken to ensure that the open pit or waste unit will meet all applicable federal or state laws, regulations and standards for air, surface water and groundwater protection following closure and [the unit] will not pose a current or future hazard to public health and safety.*[55]

Moreover, instead of outlining specific reclamation standards, the NMMA allows permit and reclamation requirements to be established on a site-specific basis, and regulations require only the use of *most appropriate technologies* and *best management practices.* New mining operations are to be designed *in a manner that incorporates measures to reduce, to the extent practicable, the formation of acid and other toxic drainage that may otherwise occur following closure, to prevent releases that cause federal or state standards to be exceeded.*[56]

New Mexico historically has allowed mining companies to pollute with few consequences, and it was one of the last western states to enact a mining reclamation law. The Molycorp molybdenum mine near Questa, on the Red River, provides an example of the lack of state regulation of hardrock mines:

Although the state had instituted a groundwater discharge permitting system 15 years earlier, it was only in November of 1992 that Molycorp was ordered to obtain a permit. In the meantime, the mine was responsible for an endless stream of mine tailings spills — 130 in 26 years.[57] Local citizens' groups have long blamed Molycorp for polluting the Red River with heavy metals. In contrast to a 1966 Public Health Department study which rated the river's chemical and biological quality as "good to excellent," the river is now considered biologically dead. And, although the mine currently is not operating because of a drop in molybdenum prices, groundwater is flooding the underground portions of the mine, risking further contamination.

PHOTO: MINERAL POLICY CENTER

Responsible for the biological death of the adjacent Red River, the Questa mine threatens further damage as its underground portions fill with groundwater.

Molycorp is not the only mine in the state with a long record of pollution in the absence of a mine reclamation program. The Chino Mine, a copper operation in southwestern New Mexico, has experienced at least 18 separate incidents of spills, leaks, and other violations between 1987 and 1992, many of which resulted in no enforcement action.[58]

The above examples document the results of hardrock mining in the absence of any reliable regulatory structure. With the implementation of the New Mexico Mining Act and reclamation regulations, there will be more oversight of mining operations than in the past. Judged by the inadequacies of the state's water quality program, however, regulation of hardrock mining in New Mexico, even with the new program, is likely to sustain continued problems.

Colorado. On paper, Colorado's hardrock and water quality regulatory programs seem to offer protection from serious degradation of the state's water resources from hardrock mining operations. Although programs are in place, however, experience has demonstrated that the state's agencies are poorly equipped — in terms of staffing, funding, and coordination — to make informed decisions and regulate hardrock operations effectively. For example, the Colorado Water Quality Control Act (WQCA) establishes the state's

Passive Treatment of Mine Drainage (PTMD) program, which addresses the construction and operation of systems to control mine drainage that is not subject to NPDES permitting. Unfortunately, however, the program does not require the construction of such systems. Furthermore, neither the WQCA nor the regulations contain any design standards (outlining specific detailed measures to be taken) applicable to mining activities. In addition, the WQCA exempts inactive and abandoned mines from its permitting requirement for surface water discharges and does not provide for groundwater permitting at all.

As a result of weaknesses in state laws and the agencies charged with carrying them out, Colorado's first large-scale cyanide heap-leaching gold mine was a disaster. Proposed in 1984 by Canadian-owned Galactic Resources and in operation from 1986 to 1992, the Summitville Mine site is high in the San Juan Mountains of southwestern Colorado, where weather conditions and steep slopes provided numerous obstacles to safe mining. Ultimately, the agencies' permitting, enforcement, and bonding efforts failed in a truly spectacular fashion. The mine caused countless cyanide spills as well as acid drainage that has significantly contributed to the eradication of aquatic life along 17 miles of the Alamosa River.[59]

The problems at Summitville stemmed from a combination of poor mine design, a hasty permitting process, inadequate bonding, ill-timed construction, and foremost a state agency that was both unable and unwilling to take action against the company until it was too late. Even a state-endorsed report admits that the regulatory climate in Colorado was "friendly to industry" and favored the mine operators throughout the decision-making process.[60] The mine closed when Summitville Consolidated Mining company filed for bankruptcy, followed shortly thereafter by a bankruptcy declaration by Galactic Resources itself. Because the state failed to require the company to post a bond at an adequate level to ensure reclamation of the site, the state had only a fraction of the funds necessary for the cleanup of Summitville. The company walked away, leaving behind a $120 million Superfund cleanup mess.[61]

As one of the worst environmental mining disasters in the nation, Summitville demonstrates clearly the inability of Colorado's regulatory agencies to protect water resources in sensitive areas from the dangers that large, modern hardrock mining operations present. The experience at Summitville caused the Colorado State Legislature to strengthen its state mining laws in 1993, particularly with respect to chemical process mines. It remains to be seen, however, whether the state will improve its regulatory performance.

Montana. Montana's hardrock and water regulatory programs have been criticized widely, both from within the state government and from outside observers. The state's regulatory agencies are plagued with problems of non-coordination and lax enforcement. In addition, Montana's system of bonding for hardrock mines is woefully inadequate, allowing operators to walk away from polluted sites, leaving only small bonds that do not begin to cover the costs of cleanup and long-term water reclamation.

Seventeen miles of the nearby Alamosa River, an irrigation source for farmland throughout the San Luis Valley, have been contaminated by cyanide and acid leaking from the Summitville heap leach gold mine.

The Montana Groundwater Pollution Control System requires a permit for operations discharging or potentially discharging pollutants to groundwater. This permit requirement does not apply, however, to mining operations with permits or licenses under the state's Metal Mine Reclamation Act; that law imposes groundwater requirements on an individual permit basis.[62] And, while the state's Water Quality Act does contain a provision for regulation of *sites disturbed by construction, modification, or operation of disposal systems* [63] and the Board of Health and Environmental Sciences may adopt rules governing reclamation of such sites and accept *voluntary* performance bonds,[64] no rules have yet been adopted.

The State of Montana has in place a policy of nondegradation of water resources. Under that policy, if a mining operation will degrade waters of the state, the operator must petition the Board of Health and Environmental Sciences for an exemption. The Board may authorize an operation to degrade state waters upon a showing of social or economic necessity or other *good cause.*

Montana's regulatory system has been criticized widely for the lack of water resource protection afforded by reclamation bonding. The system is designed to ensure surface land reclamation and short-term water treatment, and bonds are calculated with only these goals in mind. The state does not consider or provide for long-term protection of water quality. Moreover, if serious problems arise at a mine, or if a bond is forfeited, the current system only allows for money to be spent on the specific reclamation requirements outlined in the reclamation plan.

Enforcement also has been a problem for the mineral regulatory agencies in Montana. Insufficient staffing, funding problems, and lawsuit delays are blamed for the state's lack of enforcement, which is documented in a September 1994 special report of the Helena Independent Record. The report highlights a number of cases in which the Water Quality Division of the Department of Health and Environmental Science has failed to take enforcement action against violators.[65]

The State of Montana conducted legislative audits in 1994 of both hardrock mining regulation under the Hardrock Bureau and of enforcement of the Water Quality Act by the DHES. Both audits acknowledged serious deficiencies in state regulation and recommended changes to strengthen these programs. It remains to be seen whether the agencies will improve their regulatory efforts, or whether lax enforcement, underbonding, and contamination of water resources from hardrock mining will go on as before.

More Holes Than Net

Federal and state regulation as currently practiced is miserably inadequate to protect the nation's water resources from the impacts of hardrock mining. Until there is a comprehensive federal regulatory program, however, the patchwork system described here constitutes the state of regulation. It is characterized by:

- A lack of consistent and integrated regulation at the state and federal level;
- Regulatory agencies that are hampered by inadequate funding and by political pressures;
- A lack of effective monitoring and evaluation systems;
- Vague reclamation standards;
- A lack of bonding for long-term water quality impacts; and
- Exemptions which undermine the effectiveness of existing laws.

Until comprehensive reform of mining requirements at the federal level is undertaken, mining will continue to be regulated haphazardly — in an inadequate, piecemeal fashion, allowing mining activities to damage and diminish water resources.

♦ ♦ ♦

Notes

1. The Refuse Act of 1899 prohibited the disposal of "any refuse matter" into navigable waters of the United States, but its primary concern was to keep solid waste matter from blocking rivers and channels. It had little or no impact on most industrial and municipal sources of water pollution as it is known today.

2. 40 C.F.R. §122.2.

3. Based on inquiries to the Division of Oil, Gas and Mining (May 1994).

4. Personal communication with Steve McNeil, Water Quality Division (February 1995).

5. 40 C.F.R. §122.26.

6. 40 C.F.R. §130.7.

7. 40 C.F.R. §130.2 (g)-(I).

8. Idaho Sportsmen's Coalition V. Browner, U.S. District Court, Western District of Washington at Seattle, Docket No. C93-94WD, Order on Motions for Summary Judgment and Injunction (19 May 1995).

9. *Id.*

10. *Id.*

11. The body of legislation collectively referred to as RCRA includes SWDA, the 1976 RCRA, the Solid Waste Disposal Act Amendments of 1980, and the Hazardous and Solid Waste Amendments of 1984. 42 U.S.C.A. §6901.

12. 42 U.S.C.A. §6903(5).

13. 42 U.S.C.A. §6921.

14. 42 U.S.C.A. §6921(3)(A)(ii).

15. U.S. Environmental Protection Agency, Report to Congress: Wastes From the Extraction and Beneficiation of Metallic Ores, Phosphate Rock, Asbestos, Overburden from Uranium Mining, and Oil Shale (31 December 1985).

16. RCRA defines solid waste as "any garbage, refuse, sludge ... and other discarded material resulting from industrial, commercial, mining, and agricultural operations, and from community activities, but does not include solid or dissolved materials ... which are point sources subject to permits under ... the [Clean Water Act] ... " 42 U.S.C.A. §6903(27).

17. 42 U.S.C.A. §6973(a).

18. U.S. Environmental Protection Agency, Strawman II, Recommendations for a Regulatory Program for Mining Waste and Materials Under Subtitle D of the Resource Conservation and Recovery Act (21 May 1990).

19. Glenn C. Van Bever, *Mining Waste and the Resources Conservation and Recovery Act: An Overview,* Journal of Mineral Law and Policy, 7(2) (1991).

NOTES

20. NEV. CODE ANN. §444.642(3).

21. Solid waste is defined as "any garbage, trash, rubbish, refuse, sludge from a waste treatment plant or pollution control facility and *other discarded material, including solid, liquid, semisolid, or contained gaseous material* but not including domestic sewage or hazardous wastes." ARIZ. REV. STAT. §49-701.

22. ARIZ. REV. STAT. §49-851 et. seq.

23. ARIZ. REV. STAT. §49-963.

24. U. S. ENVIRONMENTAL PROTECTION AGENCY, OFFICE OF SOLID WASTE, MINING SITES ON THE NATIONAL PRIORITIES LIST: NPL SITE SUMMARY REPORTS (1991, 1996).

25. 42 U.S.C.A. §9607(4)(c).

26. 30 U.S.C.A. §21a.

27. *Id.*

28. U.S. GENERAL ACCOUNTING OFFICE, FEDERAL LAND MANAGEMENT: AN ASSESSMENT OF HARDROCK MINING DAMAGE (April 1998).

29. 43 U.S.C.A. §1732(b).

30. DEPARTMENT OF THE INTERIOR, BUREAU OF LAND MANAGEMENT, ALTERNATIVE MANAGEMENT CONTROL REVIEW OF THE MINING LAW SURFACE MANAGEMENT PROGRAM (July 1992).

31. U.S. DEPARTMENT OF THE INTERIOR, BUREAU OF LAND MANAGEMENT, SOLID MINERAL RECLAMATION HANDBOOK, BLM MANUAL HANDBOOK H-3042-1.

32. DEPARTMENT OF THE INTERIOR, BUREAU OF LAND MANAGEMENT, ALTERNATIVE MANAGEMENT CONTROL REVIEW OF THE MINING LAW SURFACE MANAGEMENT PROGRAM (July 1992).

33. *Id.*

34. *Id.*

35. *Id.*

36. *Id.*

37. Larry D. Swanson, *Federal Regulation of Hardrock Mining in the National Forests*, MONTANA BUSINESS QUARTERLY (Fall 1989).

38. 36 C.F.R. §228.4(a)

39. 43 C.F.R. §3809.3-1(c).

40. BUREAU OF LAND MANAGEMENT, AGREEMENT WITH THE STATE OF COLORADO (28 June 1984).

41. California, Colorado, Idaho, Montana, Nevada, Oregon, Utah, Washington, and Wyoming. Arizona and New Mexico were both in the process of developing state programs, with MOUs to be developed.

NOTES

42. 16 U.S.C.A. §1131.

43. U.S. DEPARTMENT OF THE INTERIOR, NATIONAL PARK SERVICE, ABANDONED MINERAL LANDS IN THE NATIONAL PARKS.

44. In addition to the states listed, California also was surveyed. However, because mine reclamation regulation is handled at the local government level by 108 cities and counties, numbers of inspectors, inspection frequencies, etc., could not be determined.

45. The Bureau has approved this many reclamation plans and has issued 173 waste discharge permits.

46. At the time these figures were given, the state had not yet adopted new regulations to implement its first mine reclamation law and create a proposed Bureau of Hardrock Reclamation.

47. This many sites had registered with the Division in compliance with the new statute; officials expected the number of known active sites to rise to about 375.

48. Environmental Law Institute, *Regulation of Mine Waste in Arizona*, STATE REGULATION OF MINING WASTE: CURRENT STATE OF THE ART, p.34 (November 1992).

49. ARIZ. REV. STAT. §49-1273.

50. ARIZ. REV. STAT. §49-1203.

51. ARIZ. REV. STAT. §49-1224(B).

52. ARIZ. REV. STAT. §49-1231.

53. N.M. STAT. ANN. §69-36-7(I).

54. NEW MEXICO MINING ACT RULES R1.1.

55. N.M. STAT. ANN. §69-36-11.B(3).

56. N.M. STAT. ANN. §69-36-7.H(5).

57. HIGH COUNTRY NEWS (22 March 1993).

58. Based on EPA and New Mexico Health and Environment Department documents, *cited in* MINERAL POLICY CENTER, MINING REPORT CARD FOR PHELPS DODGE CORPORATION (February 1993).

59. DANIELSON & MCNAMARA, THE SUMMITVILLE MINE: WHAT WENT WRONG (1993).

60. *Id.*

61. Statement by Jim Hanley, EPA Remedial Project Manager for Summitville (19 November 1993).

62. MONT. ADMIN. R. 16.20.1012(m).

63. MONT. CODE ANN. §75-5-401.

64. MONT. CODE ANN. §75-5-405.

65. HELENA INDEPENDENT RECORD (1 July 1994).

Water Pollution from Mining: An International View

9

Carlos D. Da Rosa

Carlos D. Da Rosa is Senior Research Associate at Mineral Policy Center. Prior to working for MPC, Mr. Da Rosa researched environmental issues for Representative Richard Gephardt, the Center for International Environmental Law, the Department of Justice, and the EPA. He is a graduate of Yale University and Northeastern University School of Law.

Hardrock mining occurs in most countries, projecting its impacts on water resources worldwide. Many of the effects of hardrock mining in other nations are destructive, frequently exceeding in scale the damage to water inflicted by U.S. mines. Many of these mines have carried and continue to carry out harmful practices, such as direct dumping of raw tailings into waterways, not permitted in the United States. The result has been devastation to aquatic life and harm to human populations.

Because the mining industry is active throughout the world, one cannot adequately describe its impacts on water in a short chapter. What follows is a cursory overview of mining worldwide and a description of some of the emerging trends in the regulation of mining's impacts on water in other nations. The focus will be on the developing world, where some of the worst damage continues to occur. Such a discussion is timely, because of rapidly increasing levels of foreign investment directed to these nations' hardrock mining industries.

PHOTO: PHILIP M. HOCKER/MINERAL POLICY CENTER

Mining operations may use simple equipment like pick and shovel or more modern machinery like the bulldozer shown above.

Mining in the developing world is extremely varied. Mining in these countries ranges from modern, large-scale operations — which may be government or privately owned — to small-scale, *artisanal* mining operations using simple equipment like pick and shovel, and typically run by one or a few individuals. Although large-scale operations account for most mineral production in the developing world, small and medium-scale mining operations are a significant source of mineral production and employment in many developing nations.[1] As described below, both large and small-scale mining have caused serious damage to water resources in the developing world.

However, one should not draw the inference that mining's damage to water is confined to developing nations. As demonstrated throughout this book, the United States continues to be plagued by devastating water damage from mining, past and present. Developed economies, such as Canada, Sweden, and Australia, likewise suffer from significant environmental damage from hardrock mining. Moreover, in all these nations, regulation of these harmful mining impacts is often inadequate. Thus the main difference between these nations and developing nations in terms of mining impacts and their regulation is usually a matter of degree.

Mining: A Worldwide Industry

Hardrock mining is a worldwide industry. The world value of non-fuel mining, approximately $140 billion in 1987, is expected to reach $200 billion by 2010.[2] Trade in minerals among nations is significant as well: In

1992, the world export value in minerals, ores, and metals (including coal) totaled $129 billion.[3] The United States is a major purchaser of metals, dependent upon other nations for its supplies of iron ore, bauxite (aluminum ore), cobalt, and tungsten.

Highly industrialized market economies such as the United States, Canada, Australia, and South Africa rank among the world leaders in the production of hardrock minerals. The United States is a major producer of copper, gold, phosphate rock, and a wide range of other minerals. The United States is the number one mineral producer in the world, accounting for 13 percent of the world's non-fuel mineral production.[4] South Africa, the world's greatest producer of gold, also ranks first in worldwide platinum production. Australia is a major gold and iron ore producer, and Canada is among the premier producers of zinc, gold, nickel, and uranium.

A significant portion of the world hardrock industry is also based in developing countries. Chile, for example, is the world's largest producer of copper. Other developing countries that are significant copper producers include Peru, Indonesia, Zaire, Zambia, and Papua New Guinea. Brazil, a leading producer of tin, bauxite, and iron ore, is also one of the world's top ten gold producing nations. Other significant gold producers are Indonesia, Papua New Guinea, Ghana, Zimbabwe, and Guyana. Mexico leads the world in silver production.[5]

Russia and the former Soviet republics such as Uzbekistan and Kazakhstan are also significant mineral producers of iron ore, copper, manganese, and gold. The People's Republic of China, one of the world's fastest growing economies, ranks among the world's leaders in the production of iron ore, tungsten, tin, and manganese.

Developing Nations' Dependence on Mining

Many developing nations have become highly dependent on mining as a primary means to generate revenue to run their national economies. A good measure of a country's dependence on mining is the industry's share of its total export revenues. From 1988 to 1990, for example, non-fuel minerals accounted for over 80 percent of the total value of Zaire's exports. Mineral export figures for the same period for other countries include: Papua New Guinea, 72 percent; Chile, 57 percent; Guyana, 41 percent; and Zambia, 99 percent.[6] In nations where the state owns mining operations, these operations provide important sources of revenue for the national government. Many nations, seeking to further develop their large mineral resources, are actively trying to attract more investment, capital, and exploration by mining companies based in developed countries. The developed world has responded positively.

While mining investment is surging throughout the developing world, Latin America has attracted a particular interest among outside investors. As a whole, in 1994, this region of the world accounted for the largest share of overall investment dollars into mineral "exploration" by mining companies — $544 million.

PHOTO: JOHN BURKE BURNETT

Freeport-McMoran, the New Orleans-based owner and operator of the Grasberg Mine in Indonesia (shown here), stands accused of complicity in human rights abuses and causing major environmental damage.

The 1994 figure represented a 64 percent jump in exploration spending over the previous year.[7] Indonesia, West Africa, and the former Soviet states of Kazakhstan and Uzbekistan have also become attractive areas of exploration investment from abroad. The United States, however, continues to attract a very high proportion of mineral exploration investment — ranking third in the world behind Australia and Canada.[8]

Mining Impacts in Developing Nations

Mines in developing nations often operate under very lax or non-existent regulation and enforcement. Many mines also operate in remote areas where national governments poorly monitor their operations. Even when developing countries have generated strong laws to regulate the adverse environmental impacts of mining, these laws are rarely, if ever, enforced.

Political and economic conditions in developing countries make it difficult for people to respond to cases of adverse water impacts from mining and then to communicate this information to others. Authorities in these countries, for example, often do not disclose or make available to their citizens information on environmental impacts. Sometimes, the lack of awareness or understanding of environmental problems among the general population makes it difficult to identify mines which threaten environmental quality. In other cases, governments may discourage or suppress the activities of environmental groups which focus attention on natural resource damage caused by mining. For instance, governments which are heavily dependent upon mining for their national income may have a strong incentive to conceal information about mining's harmful environmental impacts.

A further handicap in exposing mining's harms to the environment is the dearth of reliable, scientific information documenting the impacts of mining on water worldwide — a problem especially pronounced in developing nations.

As a result of these obstacles, many of the impacts that receive prominent worldwide media attention are the most dramatic and visible ones or cases in which there is considerable anecdotal evidence about water damage. Undoubtedly, there are numerous cases of mining-caused water pollution that have not yet come to international attention.

Threats to Water Resources Worldwide

Because mining in developing nations often operates under poor or nonexistent regulation, crude practices, such as the dumping of raw tailings into waterways, have flourished. These practices have resulted in metal contamination and sedimentation of waterways, causing serious environmental degradation and putting human health at risk.

Under a climate of poor regulation, the culprits of this water damage span the full range of mining operations: privately owned mines, including multinational corporations; state-owned mining corporations; and small-scale mines. Many state-owned mining operations are nationalized successors to privately owned, multinational companies that were notorious polluters. Many of these state-owned mining enterprises suffer from insufficient capital, aging equipment, and outmoded technology that contribute to the generation and inadequate treatment of water pollutants. Many small-scale operations, which often operate outside the law, commonly practice environmentally dangerous mineral processing techniques, such as mercury amalgamation, that have virtually disappeared in the U.S.[9]

Following are selected case study examples that illustrate how poor mining practices are adversely affecting water resources in developing countries:

Ok Tedi Mine, Papua New Guinea

Located in Papua New Guinea's (PNG) rainy and remote western highland region, the large Ok Tedi copper and gold mine provides one of the world's most dramatic examples of major watershed destruction caused by mining. Since the beginning of operations in 1984, the Ok Tedi mine, majority owned by the large Australian-based mining conglomerate, BHP, has dumped over 80,000 tons per day of untreated tailings directly into the Ok Tedi River, a tributary of the Fly River — one of the nation's principal waterways.[10] These tailings contain high concentrations of copper and lower concentrations of cadmium, zinc, and other contaminants, which pose serious threats to aquatic life.[11]

This mining operation has altered the course of the Ok Tedi River and destroyed most of the aquatic life along the first 70 kilometers of the river.[12] Residents along the upper Ok Tedi report that all species of fish, crocodiles, turtles and crustacea (other than some species of catfish) have been destroyed.[13] Fish *biomass* (fish life) along the upper Fly River has decreased significantly as well. The sediment load has altered the course of the Ok Tedi and caused the river to rise, flooding the rich riverside agricultural land upon which area residents have long depended.[14] Watershed damage has disrupted the traditional subsistence lifestyle of the Wopkaimin, the indigenous population living along the Ok Tedi.[15]

Environmental impacts from the Ok Tedi mine are greatest along a stretch of the lower Ok Tedi River where tailings are accumulating. There, rising floodwaters have drowned fertile community gardens and severely damaged trees, including the sago palm, the residents' most important food source. Many trees fringing the river are dying because the floodwaters have deprived their root systems of oxygen. This area of dead or dying vegetation covers 30 square kilometers, according to the mine's own estimates.[16] As of 1997, mining at Ok Tedi is expected to continue for at least another 14 years.

The Ok Tedi mine also illustrates how companies from developed nations can extract environmental concessions from developing host nations by using the country's economic weakness and vulnerability as leverage. The original agreement setting up the operation — signed in 1976 between the original private investors and the government of PNG as a minority owner — exempted the mining company from most of PNG's environmental laws.

In the early 1980s, as the mine was about to enter operation, a consortium of companies collectively known as Ok Tedi Mining Ltd (OTML) with partners including BHP and the U.S.-based Amoco, hastily built a tailings dam to contain wastes. Still under construction, the dam collapsed after a land movement. Insisting that building a tailings dam was unfeasible in this steep and unstable area, OTML prevailed upon the PNG government to allow it to discharge tailings directly into the Ok Tedi River without building a dam. The government, reluctant to discontinue an operation that provided 30 percent of the nation's total income, gave in to the company's demand.

An Australian law firm claiming to represent a group of 30,000 landowners along the Ok Tedi and Fly Rivers sued OTML and BHP in the Victoria Supreme Court in Australia in May 1994. The suit sought construction of a tailings dam, damages for loss of subsistence fisheries and fauna, compensation for harm to the traditional way of life and culture of villagers, and punitive damages. (The plaintiffs filed suit in Australia rather than in PNG because of the PNG government's close relationship to the mining operation, in which it is a 30 percent co-owner.) The suit, which sought $4 billion (Australian) in damages, was one of the world's largest environmental damages claims.

In an attempt to squelch the lawsuit, BHP and OTML negotiated an agreement with the PNG government to offer Ok Tedi and Fly River residents a compensation fund of $110 million (Australian) to pay individual damage claims made against the mine under PNG law. The original version

PHOTO: JOHN BURKE BURNETT

The dumping of untreated tailings directly into the Ok Tedi River (shown here) has destroyed aquatic life, flooded agricultural land, and displaced the local indigenous population.

of this agreement contained a provision to criminally sanction anyone who took court action against the mine in pursuit of compensation for pollution damages or loss of property.[17] (This reprehensible provision was later scrapped by the PNG Parliament when it approved the PNG-Ok Tedi agreement in December 1995.)

In June 1996, the plaintiffs, BHP, and OTML came to a settlement agreement over a compensation package. While retaining the $110 million compensation fund approved by the PNG Parliament, the agreement further provided for a special package of benefits for residents along the lower Ok Tedi River, who have been the most adversely impacted by the mine's tailings dumping. BHP also committed to "finding ways to reduce the amount of tailings entering the Ok Tedi/Fly River system." In the agreement, BHP did not pledge to manage its tailings, but stated that one of the options "under serious consideration" was piping its tailings to flat land below the mountains.[18]

Panguna Copper Mine, Papua New Guinea

The Ok Tedi mine destruction followed another infamous case of watershed damage from mining in PNG: the Panguna copper mine located in the PNG island of Bougainville. Before it closed in 1989, this major copper mine dumped 600 million tons of metal-contaminated tailings into the Kawerong River. The mine's pollution caused the death of most aquatic life in the Jaba River, into which the Kawerong River flows. The mine antagonized local landowners and, thus, played a significant role in fomenting the island's secessionist war. The war forced the mine to close down in 1989.[19] The operation has not reopened since then.

Chañaral, Chile

In Chile, the world's largest copper producer, mines have routinely deposited slag and tailings directly into rivers and the ocean despite legal prohibitions against the practice. Many poorly regulated small-time prospectors have also dumped mine wastes along streams with little concern for environmental impact. In dry areas, such as the country's mineral-rich north, dumping can be especially damaging because of the small quantity of water available to dilute contaminants. Mining in desert areas also consumes precious water supplies at the expense of other users. An observer of the Chilean mining industry notes that the nation's governing authorities have failed to develop an overarching policy to manage scarce water resources.[20]

The "Chañaral Case," named after a northern Chilean town on the Pacific coast, is one of this nation's most infamous examples of water damage from mining. In the mid-1980s, the fishing town of Chañaral took legal action against Chile's largest copper producer, the National Copper Corporation of Chile (CODELCO), for directly channeling tailings into a nearby bay. The tailings dumped by the state-owned operation killed most sea life in the area and harmed the important local fish industry. A court ruled in favor of the town of Chañaral, ordering the company to construct a tailings pond. The company has since built and begun operating a tailings pond.[21] The Chañaral case has served as an important precedent for a more recent Chilean case involving tailings dumping by an iron pellet factory in the northern city of Vallenar. The factory's long-standing dumping practices had destroyed most fish in Chapaco Bay. In response to a suit by the area's fishermen and divers, the Chilean Supreme Court ordered the factory's owner to discontinue tailings dumping into the sea.[22]

PHOTO: ANN MAEST

Untreated copper tailings flow from the Toquepala and Cuajone mines in southern Peru (on left) into the Rio Locumba (right). Where it empties into the Pacific Ocean, the Rio Locumba far exceeded U.S. aquatic life standards for copper, arsenic, and molybdenum.

Ilo, Peru

Peru, a major mineral producer, has suffered extensive environmental damage from mining. Southern Peru Copper Corporation (SPCC), majority owned by the U.S. company ASARCO, Inc., has created a significant air pollution problem in a Pacific coastal area near the southern city of Ilo. The source of the problem is SPCC's copper ore smelting operations. SPCC, which produces two thirds of Peru's copper output, has also used creek beds and a river to discharge over 94,000 tons of tailings a day from two copper mines. The tailings have accumulated in the Bahía de Ite (Ite Bay) in the form of artificial beaches. One of the world's most polluting smelters, the SPCC facility has long deposited slag — at a rate of 800,000 tons per year — into the Pacific Ocean. This smelter's dumping has created a long, black beach of mine slag along the coast.[23] The company never obtained a permit to conduct the dumping of these untreated wastes and remains in violation of Peruvian law.[24] Fishermen claim reduced catches along the coast are the result of tailings and slag dumping. Urban and agricultural residents also assert that the mines have overdrawn water supplies, depleting sources for uses other than mining.

Recent monitoring in this area by a U.S. geochemist reveals the seriousness of water contamination caused by mining. The tailings dumped by SPCC's operations eventually enter the Locumba River, which empties into the Pacific Ocean at the Bahía de Ite (see Chapter 6, Photo and Photo Insert). Total copper concentrations in the brackish water at the mouth of the Locumba River were measured at 70 μg/L — 24 times the U.S. EPA level established for the protection of marine aquatic life against chronic copper exposure, as listed in the EPA's Ambient Water Quality Criteria (AWQC). A water sample taken in the wetlands area created by the mine tailings near the river's mouth showed total arsenic concentrations at 430 μg/L — 33 times higher than the chronic marine AWQC — and total copper concentrations at 260 μ/L — 90 times higher than the chronic marine AWQC. Sands in the wetlands area, touted as a wildlife habitat area by SPCC, had copper and molybdenum concentrations as high as 27,500 mg/kg and 71 mg/kg respectively.[25]

SPCC's damage to water resources recently prompted various community groups in Ilo, Peru to present their grievances against SPCC to the International Water Tribunal, based in Amsterdam, the Netherlands. Although lacking enforcement powers, the Tribunal found SPCC guilty of negligence in contaminating rivers and coastal waters with contaminated slag and tailings. The Tribunal also concluded that the plaintiff community groups had "provided sufficient evidence that ... tailings and slag ... have harmed and diminished fish in coastal waters ... depriving [fishermen] from their major source of livelihood." The body also "regretted" that Peruvian authorities had not been able to enforce the nation's environmental legislation. SPCC, which rejected the Tribunal's jurisdiction, declined to present its case to the court.[26]

SPCC has recently been persuaded to improve some of its environmental practices. For example, the company agreed with the Peruvian government to spend $300 million in improvements in its facilities, including new funding for improved environmental technology. The company recently began constructing tailings impoundments for its two copper mines.[27]

Nonetheless, 400 citizens of Ilo, Peru, frustrated by the long lack of official action against the company, filed suit in August 1995 against SPCC in state court in Corpus Christi, Texas, where an ASARCO subsidiary is located. The plaintiffs sought compensation for health damages, mostly respiratory illnesses, from the company's smelter operations in Ilo.[28] As of early 1997, the case remains in litigation and no Peruvian court has awarded compensation to anyone adversely affected by SPCC's operations.[29]

Mantaro River Watershed, Peru

In Peru's interior, the state-owned mining giant, Centromin, has caused devastation to the Mantaro River watershed in the Andes Mountains from the mining of gold, silver, lead, copper, and other ores. Careless dumping of smelter and mine wastes into the watershed was initially started by a previous

PHOTO: HAGLER BAILLY SERVICES INC.

At the Morococha Mine (a polymetallic mine) in the Peruvian Andes, pollution from diverse mine waste sources flows into a contaminated alpine lake.

mining company. In 1973, the Peruvian government nationalized the operations and continued most of the same practices.

The watershed's contamination is acute and extensive. Total lead concentration in the Mantaro River downstream of Centromin's smelter at La Oroya has been measured at 220 μg/L, 6 times the international limit set by the World Bank.[30] Tailings disposal from the company's largest mine, located at Cerro de Pasco, has ravaged Lago de Junin, a once pristine mountain lake and national reserve, bringing the Junin grebe, a diving bird, to the brink of extinction and forcing lake-side communities to relocate more than two miles away.[31] Area residents complain that fish and wildlife have largely disappeared from the lake and other waterways and that water is unfit for drinking. Residents in La Oroya suffer from respiratory problems caused by extremely high levels of arsenic and sulfur dioxide released by Centromin's smelter (see Appendix and Chapter 7, Photo). A soil sample collected 4 kilometers downwind of the smelter contained 12,600 mg/kg of surface arsenic, 22,000 mg/kg of lead, and 305 mg/kg of cadmium.[32]

Water sampling in this region revealed high metal concentrations in 1994 resulting from acid mine drainage and other mining-related discharges —

Pictured here, a Brazilian worker operates a gold sluice.

PHOTO: ERROL D. SEHNKE, BRUCE COLEMAN, INC.

concentrations that would violate U.S. criteria for the protection of drinking water and freshwater aquatic life. Mantaro River samples taken 4 kilometers downstream of Centromin's La Oroya smelter contained total metal concentrations that significantly exceeded U.S. standards for protection of both human health and freshwater aquatic life from exposure to contaminants:

Arsenic concentrations exceeded standards by a factor of 2; cadmium by a factor of 8; copper by a factor of 16; and lead by a factor of 68.[33] Tailings water at the nearby Morococha copper mine had arsenic levels 1900 times greater than the U.S. Safe Drinking Water Act Maximum Contaminant Level (MCL), cadmium levels 270 times the MCL, and lead 4500 greater than the corresponding EPA action level.[34]

Centromin's enormous complex of copper, silver, and lead mines and smelters is now up for auction under Peru's new privatization policy. The company has had great difficulty in finding a buyer, however, because investors fear that they will have to bear the complex's staggering costs for environmental remediation. A 1993 World Bank study estimated that the company's 7 mines will require as much as $85 million in new environmental technology and that its La Oroya smelter will need an additional $38 million for solid and liquid waste treatment.[35] To make the sale offer more palatable to private investors, the Peruvian government has recently agreed to assume the costs of its company's past environmental damage.[36]

Amazon River

A gold rush in Brazil's Amazon basin threatens environmental and human health in the world's largest watershed. Beginning in the late 1970s, thousands of independent, small-scale miners rushed into Brazil's Amazon River basin, spurred by discoveries of gold in river sediments. The total number of these miners, known as *garimpeiros*, swelled to over 650,000 in the 1980s.[37]

Panning and dredging the sediment of Amazon tributaries, the miners use mercury to amalgamate the gold. The miners' excessive use of mercury during amalgamation has led to contamination of many rivers and streams.[38] Leftover mercury that does not amalgamate, or bond, with the gold in river sediments is commonly discarded in waterways together with the gold-depleted ore. Mercury vapor is also released into the atmosphere when miners heat the gold-mercury amalgam to drive off mercury and recover the gold.

During the 1980s, gold miners introduced approximately 1,000 tons of mercury into the Amazon watershed.[39] A large portion has been taken up by fish and other animals, many of which are then consumed by humans. Scientists have found alarmingly high concentrations of mercury in fish from the Madeira River, an intensively mined tributary of the Amazon River.[40] In this and other rivers, mercury concentrations have already reached dangerous levels in fish and other species; at these higher levels of the food chain, the mercury has been transformed into methyl mercury, the metals's most harmful form. Some fish samples have shown three parts per million methyl mercury, six times the internationally accepted limit.[41]

In addition, some humans who live along the Madeira River have accumulated levels of methyl mercury exceeding those which have produced the symptoms of *Minimata disease* (a devastating mercury-related disease) in Japan.[42] Hundreds of cases of mercury poisoning have been reported in the Amazon.[43]

Placer mining in Brazil has also caused severe sedimentation and siltation of streams. Tin mining in western Brazil is a primary culprit. Wildcat tin prospectors have clogged rivers with sediments during sluicing and washing of ore.[44]

Ghana

Ghana, one of Africa's most important mineral producers, has suffered poor water quality, mainly as a result of lax regulation of mining activity — which includes extensive, small-scale mining of stream beds. Many gold prospectors use mercury amalgamation despite official disapproval, and gold, bauxite, and manganese mining operations have caused severe sedimentation of waterways.[45] This is a particularly severe problem for Ghana, since most mining occurs in the high-rainfall southern region, where soil erosion is commonplace and contaminants are readily spread throughout the environment.

Emerging Water Protection Standards

As previously discussed, developing nations have generally provided less protection against the environmental damages of mining than have developed nations. The reasons are many and complex, but some of them include: an overemphasis on development to the neglect of other values; lack of funding for enforcement, environmental improvements and cleaner technology; lack of public awareness; and insufficient responsiveness by governments to popular concerns about mining's adverse environmental impacts.

Despite these problems some nations have begun to impose more stringent environmental regulations on the mining industry. Progress is slow and uneven, partly the result of great differences among developing countries in levels of economic development, political stability, education, and environmental movement.

Environmental improvement measures have taken different forms. In some cases, nations have enacted legislation which applies specifically to mining. In other cases, they have passed broad-ranging environmental legislation which applies to water pollution from mining and other sources.

Chile

Chile made significant headway in environmental protection by enacting broad environmental framework legislation in early 1994. This "Environmental Basis Law" requires environmental impact assessments for large new projects, including mines. It also provides the foundation for citizen lawsuits to enforce environmental laws. Further, the law embraces the principle of compensation for environmental damage, allowing persons harmed by water contamination from mining to sue for compensation for past damages.[46] The Eduardo Frei government in Chile has also recently announced that it will be conducting a study on the need to adopt reclamation and closure legislation for mines.

Despite such progress, the Frei government has not yet developed regulations administering the requirements for the environmental impact statements. Moreover, Chile's government has yet to implement effluent standards for many mining contaminants. Lax enforcement continues as the norm, thus undercutting the progress and intent of the new environmental regulations.

Peru

Peru's environmental framework legislation, passed in 1990, provides citizens with the right to sue in court to compel the government to enforce environmental laws. In addition, more recent environmental legislation tailored to mining activities requires environmental impact assessments for both new and expanded mining operations and requires mines to submit an annual report on each site's environmental conditions such as pollution discharges and environmental monitoring methods employed. The law also requires that mining companies submit *environmental management plans* (known as PAMAS), which are designed to achieve pollution control targets for a particular mining operation.[47]

Still, Peru's current regulation of mining is flawed, because many specific regulations and standards to carry out these broader environmental mandates have not yet been devised.[48] However, in January 1996, Peru took a step forward in environmental protection when the Ministry of Energy and Mines adopted regulations establishing maximum limits on concentrations of metals and cyanide in liquid effluents from mining operations.[49] Nevertheless, the greatest obstacle the nation must overcome in adequately protecting its water resources is its long tradition of non-enforcement of environmental regulations.

Ghana

In Ghana, the government has designed environmental and mining guidelines which require most larger new mines to submit environmental impact assessments and existing mines to submit environmental action plans.[50] In the action plans, mine operators must describe environmental conditions at individual mine sites and report on their effluent discharges and water management plans.[51] Unfortunately, the lack of effluent standards for water in the country hampers the effectiveness of these measures.[52]

In December 1994, legislation was passed in Ghana that establishes a new Ghana Environmental Protection Agency with enhanced powers to conduct environmental impact assessments and enforce environmental laws.[53]

PHOTO: WILL PATRIC, MINERAL POLICY CENTER

Three Indian spiritual leaders from the Fort Belknap Reservation lead a community march, in opposition to expanded mining activity, to the gate of the Z-L Mine in Montana (see Chapter 3, Case Study). A greater citizen voice in mining decisions is necessary to ensure protection of water resources whether in the United States or abroad.

Glimmers of Hope

In sum, modest, measured progress has occurred in the environmental regulation of mining impacts in some developing nations. This has taken place despite the efforts of many nations to lure foreign investment in mining by adopting new "mining-friendly" policies. Such enticements have taken the form of sell-offs of state-owned mining companies to private investors as well as favorable investment and tax provisions for mineral investors. These policies may place pressures on developing countries not to ratchet-up environmental regulation of mining too far, lest they frighten away foreign investors.

But better regulation alone will not be sufficient to achieve real progress in developing countries. Rather, there will be substantive achievement when national governments effectively and consistently enforce environmental regulations against mining operations. The best way to ensure such enforcement is for citizens of developing nations to have a greater voice in decision-making on environmental and natural resources issues. This enhanced citizen role will be pivotal in guaranteeing that governments carry out their professed commitment to improved environmental regulation of the mining industry. Enhancing citizens' rights in the field of mining regulation, whether in developing countries or in the United States, can only strengthen the regulatory process and serve as a critically important tool in efforts to protect water from mining pollution.

Mining ventures, as economically attractive as they may appear, may last only 10 to 20 years before the ore is played out. Communities, on the other hand, have lived in many mining regions for generations before mining appeared and will do so long after the mining operations have ceased. But during the relatively short life-span of a mine, the region's water resources can be severely damaged, leaving behind contaminating effects that can last for generations, continuously threatening a community's long-term well-being.

Therefore, the best way to protect the world's water resources from irresponsible mining is to give people the right and the ability to participate in the development and oversight of mining activities.

♦ ♦ ♦

Notes

1. ALYSON WARHURST, ENVIRONMENTAL DEGRADATION FROM MINING AND MINERAL PROCESSING IN DEVELOPING COUNTRIES, p.56 (Organization for Economic Cooperation and Development, 1994).

2. BARRY MAMADOU, ECONOMIC SIGNIFICANCE OF MINING AND MINERAL MARKET TRENDS, p.2 (Presented at the International Conference on Development, Environment, and Mining, Washington, D.C.) (1 June 1994).

3. UNITED NATIONS CONFERENCE ON TRADE AND DEVELOPMENT (UNCTAD), UNCTAD COMMODITY YEARBOOK 1995, p.38 (1995).

4. FACT SHEET (UNDATED), *citing* UNITED NATIONS CONFERENCE ON TRADE AND DEVELOPMENT (UNCTAD) (1990).

5. METALS & MINERALS ANNUAL REVIEW (Mining Journal Ltd., 1996).

6. FACT SHEET, *supra* note 4.

7. METALS ECONOMICS GROUP, PRESS RELEASE (14 October 1994).

8. C.J. Moon et al., *Mineral Exploration,* MINING ANNUAL REVIEW, p.4 (1996).

9. WARHURST, *supra* note 1, pp.46-49.

10. HELEN ROSENBAUM & MICHAEL KROCKENBERGER, AUSTRALIAN CONSERVATION FOUNDATION, REPORT ON THE IMPACTS OF THE OK TEDI MINE, p.i (November, 1993).

11. *Id.*

12. *Id.*

13. SLATER & GORDON, BARRISTERS & SOLICITORS, PRESS RELEASE, p.2 (5 January 1995).

14. ROSENBAUM & KROCKENBERGER, *supra* note 10, p.i.

15. David Hyndman, *Ok Tedi: New Guinea's Disaster Mine,* THE ECOLOGIST, Vol. 18, No.1, pp.24-29 (1988).

16. OK TEDI MINING LIMITED, ANNUAL REVIEW 1995, p.12 (1996).

17. Mary-Louise O'Callaghan and Matthew Stevens, *BHP Deal to Bar Ok Tedi Compo Cases,* THE AUSTRALIAN (undated).

18. BHP, NEWS RELEASE (11 June 1996).

19. John Young, *Mining the Earth, in* STATE OF THE WORLD, p.107 (1992).

20. GUSTAVO LAGOS, INSTRUMENTOS REGULATORIOS Y ECONÓMICOS PARA LA GESTIÓN AMBIENTAL DE LOS RECURSOS MINEROS: EL CASO DE LA PEQUEÑA Y MEDIANA MINERÍA [REGULATORY AND ECONOMIC INSTRUMENTS AFFECTING THE ENVIRONMENTAL PERFORMANCE OF THE MINING INDUSTRY: THE CASE OF SMALL AND MEDIUM-SCALE MINING], p.39 (December, 1993).

21. GUSTAVO LAGOS, MINING AND THE ENVIRONMENT IN CHILE (Working Paper 92-7, Colorado School of Mines), p.8 (October, 1992).

22. *Id.*

23. Sally Bowen, *Business and the Environment: Mines Make a Clean Sweep-Peru's Biggest Copper Producer is Taking the Lead in Reducing Industrial Pollution,* THE FINANCIAL TIMES, 4 August 1993, p.7.

24. DORIS BALVÍN DÍAZ, AGUA, MINERÍA Y CONTAMINACIÓN: EL CASO SOUTHERN PERU [WATER, MINING, AND CONTAMINATION: THE SOUTHERN PERU CASE], pp.170-241 (1995).

25. Data from samples taken in area of Ilo, Peru in February 1994. Samples taken by Ann Maest, geochemist, RCG/Hagler, Bailly, Inc., Boulder, Colorado. Described in Memo from Ann Maest and Dick Kamp, Border Ecology Project, to Peruvian Mayors (19 October 1994).

26. VERDICT OF THE INTERNATIONAL WATER TRIBUNAL (18 February 1992).

27. Bowen, *supra* note 23.

28. HILLIARD & MUÑOZ, PRESS RELEASE (September, 1995).

29. Telephone communication with Andrew Schirrmeister, Senior Partner, Zummo & Schirrmeister, Houston, Texas (25 February 1997).

30. Corinne Schmidt, *How Brown Was My Valley,* NEWSWEEK, 18 April 1994, p.21.

31. *Id.*

32. Data from samples taken in area of Ilo, Peru in February 1994. Samples taken by Ann Maest, geochemist, RCG/Hagler, Bailly, Inc., Boulder, Colorado. Described in Memo from Ann Maest and Dick Kamp, Border Ecology Project, to Peruvian Mayors (19 October 1994).

33. Data are from La Oroya area water samples taken during February 1994. Samples taken by Ann Maest, geochemist, RCG/Hagler, Bailly, Inc., Boulder, Colorado. Described in Memo from Ann Maest and Dick Kamp, Border Ecology Project, to Peruvian Mayors (19 October 1994).

34. *Id.*

35. *Centromin Buyer Faces Huge Environmental Investment,* THE ANDEAN REPORT, 21 February 1994, p.1.

36. *Centromin's Second Sell-off Attempt...,* MINING JOURNAL, p.310 (28 April 1995).

37. Julia Preston, *Gold Rush Brings Mercury Poisoning to Amazon,* THE WASHINGTON POST, 17 February 1995, p.A31.

38. WARHURST, *supra* note 1. p.29.

39. Preston, *supra* note 37, p.A31.

40. Olof Malm et al., *Mercury Pollution Due to Gold Mining in the Madeira River Basin,* AMBIO, February 1990, p.11.

41. Preston, *supra* note 37, p.A31.

42. "Minimata" disease is caused by ingestion of methyl mercury. The symptoms include tremors, lack of muscular coordination, deafness, comas, and death. The disease is named after the Japanese town where the most famous case of mercury poisoning occurred. Fifty people there died in the 1950s and early 1960s after becoming contaminated by eating fish containing high levels of methyl mercury.

43. Preston, *supra* note 37, p.A31.

44. Julia Preston, *Brazilian Prospectors Confront Mining Firms,* THE WASHINGTON POST, 12 November 1994, p.C4.

45. HENDRIK G. VAN OSS, U.S BUREAU OF MINES, THE MINERAL INDUSTRY OF GHANA-1993, MINERAL INDUSTRIES OF AFRICA, MINERALS YEARBOOK VOLUME III, p.57 (1994).

46. Lagos, *supra* note 21, pp.19-20.

47. KATE HARCOURT & LILIANA UNTEN, THE DEVELOPMENT OF ENVIRONMENTAL CONTROLS IN THE MINING INDUSTRY IN PERU, pp.4-5 (1994).

48. *Id.*

49. *New Effluent Standards Established in Peru,* ENGINEERING AND MINING JOURNAL, April 1996, pp.14-15.

50. Van Oss, *supra* note 45, p.57.

51. SEMINAR ON THE EFFECT OF MINING ON GHANA'S ENVIRONMENT WITH PARTICULAR REFERENCE TO PROPOSED MINING ENVIRONMENTAL GUIDELINES pp.135-139 (Peter C. Acquah, ed., 1992)

52. UNITED NATIONS CONFERENCE ON TRADE AND DEVELOPMENT, NATURAL RESOURCES MANAGEMENT AND SUSTAINABLE DEVELOPMENT: THE CASE OF THE GOLD SECTOR IN GHANA, p.35 (15 August 1995).

53. Telephone communication with Hendrik G. Van Oss, U.S. Bureau of Mines (26 June 1995).

♦ ♦ ♦

Appendix: Toxicity of Selected Mining Contaminants

Chemical exposures may be considered toxic to human health if they produce harmful effects or if they result in death. Paracelsus, a 15th century toxicologist, first expressed the key concept in toxicology as follows: "All substances are poisons; there is none which is not a poison. The right dose differentiates a poison and a remedy." Thus, the term "toxic chemical" is redundant: All chemicals are toxic if the dose is high enough. Many chemicals, such as drugs, are beneficial at lower doses and toxic at higher doses. Among chemicals, there is a wide spectrum of doses capable of producing adverse effects, injury, or death.

Whether a chemical elicits toxicity depends on several factors:

- Route of exposure (breathing, eating, touching).
- Duration and frequency of exposure (once, sometimes, every day).
- Amount of chemical in the medium of exposure (air, water, soil).
- Other factors, such as the susceptibility of the people exposed.

The purpose of this section is to identify those chemicals that might produce adverse effects as a result of exposure to mining wastes and to identify the exposure levels at which those effects might occur. The following is a summary of the toxicity of each of the contaminants that predominate in mining wastes and the doses, exposure durations, and routes of exposure that have been reported to elicit toxicity (see also Chapter 3, National Water Quality Standards for Selected Mining Contaminants).

Arsenic

Arsenic occurs naturally in many kinds of rock, especially in ores that also contain copper and lead. The primary sources of arsenic contamination in the environment are metal smelting, chemical manufacturing, pesticide application, and coal combustion. Food is the primary source of human exposure to arsenic, although it also occurs in cigarette smoke.

Arsenic has been recognized as a human poison since ancient times, with large oral doses producing death from fluid loss and circulatory collapse. Smaller oral doses produce gastrointestinal pain, hemorrhage, nausea, vomiting, diarrhea, anemia, and neurologic toxicity such as headache, lethargy, confusion, hallucination, seizures, and coma. Long-term, low-level oral exposure to arsenic may result in cardiovascular toxicity, anemia, liver toxicity, and a pattern of skin changes that includes darkening of the skin and the appearance of small corns or warts.

Skin cancer has been associated with long-term, low-level exposure to arsenic through drinking water, and there is suggestive evidence of increased risks of bladder, kidney, liver, and lung tumors as well. Long-term inhalation exposure to arsenic can lead to respiratory tract toxicity, gastrointestinal toxicity, cardiovascular toxicity, and neurotoxicity. It has also been associated with increased rates of lung cancer, especially at or near copper smelters. There is some evidence that women who had been exposed to arsenic-containing dusts where they worked were more likely to have spontaneous abortions or malformed babies than women who were not exposed. Skin contact with arsenic can cause allergic dermatitis.[1]

The International Agency for Research on Cancer and the EPA classify arsenic as a known human *carcinogen* (an agent producing or inciting cancer). The EPA restricts the level of arsenic in drinking water to 50 $\mu g/L$ (micrograms per liter), based on its carcinogenicity, and taking into consideration its potential nutrient requirements. This standard is currently under review by the EPA which could result in a more restricted standard. Most drinking water has arsenic levels of about 2.5 $\mu g/L$. The EPA also considers arsenic a *Hazardous Air Pollutant.* Various states have guidelines for the level of arsenic in air that range from 2.0×10^{-4} to 6.7×10^{-1} $\mu g/m^3$ (micrograms per cubic meter of air) on an annual average basis.[2,3] (In urban areas, arsenic levels normally range up to 3×10^{-5} $\mu g/m^3$.)

Cadmium

Cadmium is naturally distributed in trace amounts throughout the environment, but most cadmium in the environment has been released from human activities, including mining. Commercially, cadmium and cadmium compounds are widely used in products such as metal alloys, batteries, jewelry, fluorescent lamps, and pesticides. Most people are exposed to cadmium from food, especially shellfish, and from cigarette smoke.

Short-term inhalation of cadmium dust or fumes in the workplace has been reported to cause respiratory tract irritation, lung toxicity, and bronchitis. Long-term inhalation of lower levels of airborne cadmium can produce decreased respiratory function, emphysema, and lung cancer. Both oral and inhalation exposure to cadmium over long periods of time can lead to cadmium accumulation in the kidneys, causing severe kidney damage. Heavy smoking has also been reported to considerably increase tissue cadmium levels. Short-term oral exposures to high levels of cadmium can lead to gastrointestinal irritation, nausea, vomiting, and pain.[4]

The EPA considers cadmium a probable human carcinogen when it is inhaled. According to EPA guidelines, the level of cadmium in drinking water is not permitted to exceed 5 μg/L, based on an estimate of the concentration that might be associated with kidney accumulation and toxicity. In general, drinking water cadmium levels average about 3 μg/L. Lifetime oral exposure to cadmium from all sources should not exceed 100 μg/day. The federal Occupational Safety and Health Administration (OSHA) does not permit average concentrations of cadmium in workplace air to exceed 5 $\mu g/m^3$.

Chromium

Chromium occurs in the environment in several chemical forms. Elemental chromium does not occur naturally. The primary form of chromium in the environment is called *trivalent chromium,* an essential nutrient that is found as a trace element in many foods. The other form of chromium, *hexavalent chromium,* is relatively unstable and rarely occurs naturally, but can be introduced to the environment as a result of human activities, such as mining. Chromium speciation (species formation) in water depends on characteristics such as pH levels. Hexavalent chromium predominates (builds power) under the conditions generally found in shallow aquifers, and trivalent chromium predominates under the conditions in deeper groundwater.

Trivalent chromium plays an important, although not fully understood, role in metabolizing carbohydrates in mammals by forming a complex with insulin to metabolize the carbohydrates. Chromium deficiency is characterized in humans by glucose intolerance. The minimum dietary intake of chromium necessary to maintain good health is not known, although the National Academy of Sciences recommends a daily intake of 50 to 200 μg.[5] Trivalent chromium is practically nontoxic, which is partly due to its poor solubility and gastrointestinal absorption.

By contrast, hexavalent chromium compounds have been reported to be carcinogenic in both humans and laboratory animals. Three of four *epidemiologic studies* (examining the incidence, distribution, and control of disease in a population) of chromate workers reported significant excesses of lung or respiratory tract cancer, although none was controlled for potential factors, such as smoking or other occupational exposures associated with lung cancer. Inhalation of hexavalent chromium by chromate workers has also been reported to produce nasal septum perforation, asthma, irritation, and other signs of respiratory distress, and it may increase mortality from non-cancer respiratory disease. Long-term oral exposure to low levels of chromium compounds has been associated with gastrointestinal disorders. There is some evidence that hexavalent chromium produces developmental and reproductive effects in mice exposed orally during gestation.

The EPA considers hexavalent, but not trivalent, chromium compounds to be known human carcinogens following inhalation exposure. For oral exposure, EPA guidelines limit hexavalent chromium exposure to about 350 μg/day and trivalent chromium exposure to 70,000 μg/day, based on studies in which no effects were observed after oral administration to rats. Most chromium exposure comes from food, ranging between 25 and 250 μg/day of trivalent chromium. The level of total chromium in drinking water is limited by the EPA to 100 μg/L, but is usually less than 2 μg/L. OSHA recommends that the average concentration of trivalent chromium to which workers are exposed not exceed 50 $\mu g/m^3$ air.

Copper

Copper is an essential dietary nutrient in humans and animals. Copper and its compounds are naturally present in the earth's crust, and natural discharges to air and water, such as windblown dust and volcanic eruptions, may be significant sources of environmental copper. The largest anthropogenic (human related) sources of copper are mining, smelting, agriculture, solid waste, and sludge disposal. The chemical form of copper is an important determinant of its behavior in the environment.[6]

Factory workers exposed for long periods of time to high levels of copper dust in the air have experienced irritation of the mouth, eyes, and nose. Factory workers have also reported anorexia, nausea, occasional diarrhea, headache, dizziness, and drowsiness following long-term copper dust exposure. Gastrointestinal, liver, and kidney effects have been observed in humans ingesting high dosages of copper sulfate in attempted suicides. The gastrointestinal effects included vomiting, diarrhea, nausea, abdominal pain, and a metallic taste in the mouth. Some liver and kidney toxicity has also been reported in cases of infant poisoning. Adverse health effects from copper deficiency are rare in the United States, but can include anemia, poor growth, and central nervous system effects.

Experiments in laboratory animals have reported mild respiratory irritant effects in mice and hamsters exposed to airborne copper sulfate, and liver and kidney effects in rats and pigs ingesting copper.

The EPA restricts copper in drinking water to a level of 1,300 μg/L, based on gastrointestinal irritation. Copper levels in drinking water average about 75 μg/L, but may be higher if copper plumbing is used. There are no laws regulating the level of copper in air, but OSHA recommends that copper exposure in workplace air not exceed an average level of 100 $\mu g/m^3$ for copper fumes and 1,000 $\mu g/m^3$ for copper dusts and mists. The National Academy of Sciences recommends a daily intake of dietary copper of 2 to 3 mg.[7]

Cyanide

Cyanide can occur naturally in the environment, primarily in plants, but most of the cyanide in the environment results from human activities.[8] Cyanide discharged into water comes primarily from metal finishing industries, iron and steel mills, and organic chemical industries. Cyanide is also increasingly used in gold mining processing, as discussed in Chapter 2. However, road and agricultural runoff can also be a source. Most of the cyanide emitted to air comes from automobile exhaust. People can be exposed to cyanide if they eat plants that contain it naturally (e.g., lima beans, almonds, sweet potatoes, spinach), breathe contaminated air, smoke cigarettes, or drink contaminated water. In the body, cyanide is a component of vitamin B_{12}, an essential component of the diet.

The severity of cyanide toxicity depends on its chemical form. In water, cyanide is present primarily as hydrogen cyanide, although it may also be present as a metal complex. Cyanides in water are often biodegraded by bacteria. In air, cyanide occurs primarily as hydrogen cyanide gas.

Short-term exposures to high levels of cyanide compounds, whether by inhalation, ingestion, or dermal exposure, can be very toxic, leading ultimately to death. Exposure to cyanide compounds can also cause breathing irregularities, central nervous system toxicity, and gastrointestinal corrosion in both humans and laboratory animals. Because it is highly toxic, cyanide has often been used with suicidal or homicidal intent and as a chemical warfare agent. Long-term exposure to cyanide in workplace air has been associated with decreased respiratory function, cardiac pain, vomiting, headaches, and effects on the thyroid gland. Long-term oral exposure to lower levels of cyanide, such as eating a diet high in foods that contain cyanide, can cause decreased thyroid function and central nervous system toxicity in humans and animals, as well as developmental abnormalities in the offspring of laboratory animals.

The EPA has proposed limiting exposure to cyanide in drinking water to 200 $\mu g/L$ (based on weight loss observed in a two-year oral study of rats). Cyanide occurs in most drinking water supplies at levels of 0.5 to 0.8 $\mu g/L$. The EPA also recommends that lifetime oral exposure to all sources of cyanide compounds should not exceed levels of between 1400 and 14,000 μg/day, depending on the compound. OSHA restricts cyanide exposure in workplace air to an average concentration of 5,000 $\mu g/m^3$.

Lead

Lead occurs naturally in the environment, although most of the lead dispersed throughout the environment results from human activities. Combustion of leaded gasoline is the primary source of human exposure to lead in areas where it is still used. Other important sources of exposure to lead include living in urban areas or near metal smelters, smoking, and occupational exposure.[9]

Children are at the highest risk of adverse health effects from lead exposure. When children are exposed to lead either before or after birth, normal growth and development may be affected, leading to deficiencies in height, weight, intelligence, and attention. Children who live near copper smelters have blood lead levels that are significantly higher than children who do not. Even higher blood lead levels are measured in children who live near copper smelters and whose fathers are employed there.[10, 11, 12]

In adults, lead can be toxic to the central nervous system, producing weakness in the fingers, wrists, and ankles. It can increase blood pressure and cause anemia, and it can damage the kidneys of both children and adults. Lead may increase the rate of spontaneous abortion and damage the male reproductive system, leading to sterility.

The EPA's Ambient Air Quality Standard for lead is 1.5 $\mu g/m^3$, averaged over three months, and the level of lead permitted in drinking water is 15 $\mu g/L$. Most drinking water supplies have lead levels less than 5 μg/liter. The U.S. Center for Disease Control recommends that blood lead levels of 10 to 15 μg/deciliter should not be exceeded in children. It is not possible to determine the exposure levels of lead associated with these blood levels because of the complex way in which lead interacts with the body. The EPA believes there may be no threshold, or safe exposure level, for lead exposure and intelligence deficits in children.[13]

Mercury

Mercury in the environment comes naturally from the earth, but also from industrial activities, such as mining and smelting.[14] Mercury in air results primarily from natural degassing of mineral forms of mercury from rocks, but it can also result from mining and smelting activities, incineration, and coal-fired power plants. Mercury can occur in soil from the direct application of mercury-containing fertilizers, lime, and fungicides. Mercury is released to water from rocks, but it may also be released into water in wastewater effluents from industrial plants that use mercury. Common uses of mercury include batteries, thermometers, fungicides, antiseptics, and dental fillings. Most human exposure occurs as a result of eating food contaminated with mercury, usually fish or shellfish, and from the mercury vapor released from dental fillings.

The toxicity of mercury depends on its chemical form. The form of mercury found in fish and shellfish is *organic*, or carbon-containing, mercury. The form of mercury found in dental fillings is *metallic mercury*. And the form of mercury associated with rocks and with mining and smelting is *inorganic*, or noncarbon-containing, mercury. Inorganic mercury in the environment can be transformed to organic mercury by microorganisms. The most common form of organic mercury, *methylmercury*, can accumulate in some types of fish to high concentrations. Human exposure to methylmercury occurs when contaminated fish or shellfish are eaten. Methylmercury is more likely to cause nervous system toxicity than inorganic mercury.

Both short- and long-term oral exposure to inorganic mercury salts can lead to kidney damage, including kidney failure. Oral ingestion of inorganic mercury is very irritating to the gastrointestinal tract and can cause nausea, vomiting, pain, ulceration, and diarrhea. Toxicity to the brain and nervous system has been reported following large doses of inorganic mercury taken medicinally. Kidney and gastrointestinal toxicity has also been observed in laboratory animals administered inorganic mercury compounds orally. It is not known whether inhalation exposure to inorganic mercury compounds is toxic.

EPA guidelines restrict the level of mercury in drinking water to 2 μg/L, although mercury levels in water do not usually exceed 0.025 μg/L. The U.S. Food and Drug Administration bans consumption of methylmercury-contaminated fish and shellfish when it determines that the mercury levels are unhealthful. The EPA considers mercury a Hazardous Air Pollutant and restricts the amount of mercury that can be emitted from mercury ore processing plants, battery plants, and sludge incineration plants. OSHA does not permit concentrations of inorganic mercury vapors to exceed 100 μg/m3 in workplace air.

Nickel

Nickel compounds and alloys have been in commercial use for more than 100 years. Many nickel compounds are water-soluble and occur naturally in soil, water, and foods. Water-insoluble nickel compounds occur in certain industries, and human exposures can occur by inhaling dusts or fumes in the workplace. The primary source of nickel in the atmosphere is the burning of fossil fuels, and the primary source of nickel in water is from industrial pollution and waste disposal.[15] Nickel also enters the atmosphere in emissions from mining and refining operations. Plants can accumulate nickel from soil, and the diet is estimated to contribute 90% of human daily exposure to nickel. Nickel also occurs in cigarette smoke. The biologic effects of nickel compounds vary widely, so it is important to identify the chemical form of nickel that is of concern.

Some evidence supports the potential carcinogenicity to humans and animals of several nickel compounds under certain exposure conditions. In humans, respiratory-tract tumors have been reported among workers inhaling nickel refinery dust or nickel subsulfide in the workplace. Workers who inhale metals, including nickel compounds, have developed other respiratory effects, such as chronic bronchitis, asthma, and emphysema. Animal studies have also produced some respiratory tract effects, including cancer, in rats exposed to nickel compounds by inhalation. While nickel compounds have not been shown to be carcinogenic in either humans or animals following ingestion, gastrointestinal distress has been reported in workers who drank water contaminated with high levels of nickel. Lower levels of nickel are consumed daily in the diet; daily consumption of up to 2 mg is considered normal. Many people are sensitive to skin contact with nickel, such as from jewelry or coins, and they can develop dermatitis (skin rashes). Nickel is known to be essential to maintain the health of animals, and it may also be required by humans.

The EPA considers nickel refinery dust and nickel subsulfide to be human carcinogens when inhaled. The EPA has proposed a limit for soluble salts of nickel in drinking water of 100 μg/L (as nickel), based on reduced body and liver weights reported in a two-year dietary study using rats. Most of the nickel in drinking water comes from water distribution systems, and is generally less than 10 μg/L. The EPA also recommends that lifetime oral exposure to all sources of water-soluble nickel compounds not exceed a level of 1,400 μg/day. The OSHA restricts workplace exposure to metallic nickel and water-insoluble nickel compounds to 1,000 $\mu g/m^3$ in air, and water-soluble nickel compounds to 100 $\mu g/m^3$.

Zinc

Zinc, an essential dietary nutrient, is also one of the most widely-used metals in the world.[16] Zinc compounds are found naturally in air, soil, water, and food. Zinc is present in the earth's crust and can also be released to the environment from metallurgic waste related to mining, smelting, and refining. Zinc is found as a natural mineral in many drinking waters. In fact, a major source of human exposure to zinc can be drinking water that contains zinc as a result of soil erosion and corrosion of plumbing systems that use zinc-containing materials. However, the primary source of human exposure to zinc is food. Zinc has been detected in all food classes tested, with meats and seafood containing relatively high concentrations compared to fruits and vegetables. Zinc compounds are also permitted for use as food additives, as color additives in drugs and cosmetics, and as medical treatments for problems, such as acne and sickle-cell anemia.

Inadequate dietary zinc intake can lead to appetite loss, poor growth and development, birth defects, slow wound healing, and skin lesions. Too much zinc (at least ten times the recommended daily dose) can produce gastrointestinal disturbances, such as pain, cramping, nausea and vomiting, diarrhea, and pancreatic toxicity. Long-term ingestion of zinc compounds at lower doses has led to copper deficiency by interfering with the body's ability to take in and use copper. Gastrointestinal effects and copper deficiency have also been produced in laboratory animals administered oral doses of zinc compounds. Short-term exposures to high levels of zinc dust or fumes in the workplace can cause impaired lung function and respiratory irritation — effects that are reversible if exposure ceases.

The EPA restricts zinc in drinking water to a suggested level of 5,000 μg/L, based on taste. Normally, zinc levels in drinking water range from 100 to 1,000 μg/L, much of which is thought to come from plumbing systems. There are no laws regulating the level of zinc in air, but OSHA recommends that zinc chloride exposure in workplace air should not exceed an average level of 1,000 $\mu g/m^3$, while the National Institute for Occupational Safety and Health recommends that levels of zinc oxide should not exceed 5,000 $\mu g/m^3$ in workplace air. The National Academy of Sciences recommends a daily intake of dietary zinc of 15 mg for men and 12 mg for women.[17]

Notes

1. AGENCY FOR TOXIC SUBSTANCES AND DISEASE REGISTRY (ATSDR), U.S. DEPARTMENT OF HEALTH AND HUMAN SERVICES, TOXICOLOGICAL PROFILE FOR ARSENIC, *TP-92/02* (1992).

2. *Id.*

3. Integrated Risk Information System (IRIS), On-line data base (1995).

4. AGENCY FOR TOXIC SUBSTANCES AND DISEASE REGISTRY (ATSDR), U.S. DEPARTMENT OF HEALTH AND HUMAN SERVICES, TOXICOLOGICAL PROFILE FOR CADMIUM, TP-92/02 (1992).

5. NATIONAL RESEARCH COUNCIL (NRC), RECOMMENDED DAILY ALLOWANCES, TENTH ED., NATIONAL ACADEMY PRESS, WASHINGTON, D.C. (1989).

6. AGENCY FOR TOXIC SUBSTANCES AND DISEASE REGISTRY (ATSDR), U.S. DEPARTMENT OF HEALTH AND HUMAN SERVICES, TOXICOLOGICAL PROFILE FOR COPPER, TP-90-08 (1990).

7. NATIONAL RESEARCH COUNCIL, *supra* note 5.

8. AGENCY FOR TOXIC SUBSTANCES AND DISEASE REGISTRY (ATSDR), TOXICOLOGICAL PROFILE FOR CYANIDE, TP-92/09, U.S. DEPARTMENT OF HEALTH AND HUMAN SERVICES, WASHINGTON, D.C. (1992).

9. AGENCY FOR TOXIC SUBSTANCES AND DISEASE REGISTRY (ATSDR), U.S. DEPARTMENT OF HEALTH AND HUMAN SERVICES, TOXICOLOGICAL PROFILE FOR LEAD, TP 92/12 (1992).

10. D. Gagné , & G. Létourneau, *Determining soil contamination patterns in a residual district near a copper smelter,* ARCHIVES ENVIRONMENTAL HEALTH, (Volume 48), pp.181-183 (1993).

11. L. Chenard, F. Turcotte, & S. Cordier, *Lead absorption by children living near a primary copper smelter,* CANADIAN JOURNAL PUBLIC HEALTH, 78, pp.295-297 (1987).

12. D.E. Morton, A.J. Saah, S.L. Silberg, W.L. Owens, M.A. Roberts, & M.D. Saah, Lead absorption in children of employees in a lead-related industry, AMERICAN JOURNAL OF EPIDEMIOLOGY, 115, pp.549-555 (1982).

13. Integrated Risk Information System (IRIS), *supra* note 3.

14. AGENCY FOR TOXIC SUBSTANCES AND DISEASE REGISTRY (ATSDR), U.S. DEPARTMENT OF HEALTH AND HUMAN SERVICES, TOXICOLOGICAL PROFILE FOR MERCURY (UPDATE), TP-93/10 (1993).

15. AGENCY FOR TOXIC SUBSTANCES AND DISEASE REGISTRY (ATSDR), U.S. DEPARTMENT OF HEALTH AND HUMAN SERVICES, TOXICOLOGICAL PROFILE FOR NICKEL, TP-92/14 (1992).

16. Agency for Toxic Substances and Disease Registry (ATSDR), U.S. Department of Health and Human Services, Toxicological Profile for Zinc (1989).

17. National Research Council, *supra* note 5.

♦ ♦ ♦

INDEX

Bibliography of Suggested Reading

Acid Mine Drainage: Designing for Closure. John W. Gadsby, et al., Editors. (Papers presented at the GAC/MAC Joint Annual Meeting, Vancouver, B.C., Canada, 16-18 May 1990). Vancouver: BiTech Publishers Ltd., 1990.

These collected papers, authored by technical experts, discuss methods of mine planning, waste sampling, waste management, and reclamation that operators can use to prevent acid mine drainage upon mine closure.

Balvín Diaz, Doris. *Agua, Minería y Contaminación: El Caso Southern Peru. [Water, Mining and Contamination: The Southern Peru Case].* Ilo, Peru: Ediciones Labor, 1995.

This Spanish-language book examines the severe water contamination caused by the operations of the U.S.-owned Southern Peru Copper Corporation in southern Peru.

Cyanide and the Environment. Two Volumes. Dirk Van Zyl, Editor. (Collected proceedings of a conference held in Tucson, AZ, 11-14 December 1984). Fort Collins, CO: Colorado State University, 1985.

This collection of papers discusses the chemistry, toxicity, and environmental impacts of cyanide compounds, the role of cyanide in processing gold and silver, the environmental regulation of cyanide, and methods of treating cyanide in mine wastes.

Introduction to Evaluation, Design, and Operation of Precious Metal Heap Leaching Projects. Dirk J.A. van Zyl, et. al., Editors. Littleton, CO: Society of Mining Engineers, Inc., 1988.

This book provides an introduction to cyanide heap leaching, the technology that has made the United States the number two gold producing country in the world, by allowing the recovery of gold from extremely low-grade ores. The collected papers in this book discuss designing, building, and lining leach heaps, as well as some environmental considerations, such as detoxifying cyanide.

Kelly, Martyn. *Mining and the Freshwater Environment.* London: Elsevier Applied Science, 1988.

This book surveys the effects on freshwater plants and animals of heavy metal contamination and acid mine drainage caused by mining.

Leshy, John D. *The Mining Law.* Washington, D.C.: Resources for the Future, 1987.

This book examines the 1872 Mining Law, the antiquated law which still provides the framework and stimulus for environmentally destructive mining on publicly-owned, federal land in the western United States. Describing the Law's origins and implementation, the author discusses how the law fails to address the public's concerns about mining's harms to natural resources.

Lyon, James S., et al., *Burden of Gilt.* Washington, D.C.: Mineral Policy Center, June 1993.

This Mineral Policy Center report surveys the nationwide problem of abandoned mines. The report estimates that there are over 557,000 abandoned mines in the United States, 15,000 of which pose a serious risk to surface and groundwater. Mineral Policy Center estimates that cleanup of U.S. abandoned mines may range from $32 billion to $72 billion.

McElfish, James M., Jr., et al. *Hard Rock Mining: State Approaches to Environmental Protection.* Washington, D.C.: Environmental Law Institute, 1996.

This book describes the environmental regulation of hardrock mining by U.S. states, which today play the leading role in regulating the industry's operations and environmental impacts. Examining mining laws and regulations of seven western U.S. states, the authors explain how these states address such issues as water protection, operating and reclamation standards, inspections, and enforcement.

Mine Waste Management. Ian P.G. Hutchison & Richard D. Ellison, Editors. (Sponsored by California Mining Association). Boca Raton, FL: Lewis Publishers, 1992.

This book provides a mining industry perspective on design and operating techniques that are effective in preventing water contamination from mining. The book discusses how to predict acid mine drainage, how to incorporate climactic considerations in mine design, and how to line tailings and heap leach facilities to prevent groundwater contamination.

The Mining Waste Study Team of the University of California. *Mining Waste Study.* Berkeley: University of California, 1 July 1988.

This report, commissioned by the California State Legislature, examines the environmental and public health risks posed by hardrock mining in

California, and proposes improvements in regulating mining's environmental impacts as well as techniques to better predict and prevent mining-caused water contamination.

Rickard, T.A. *History of American Mining.* New York: McGraw-Hill, 1932.
This book describes the history of early U.S. mining, from the pre-colonial era to the early twentieth century, with an emphasis on mining booms of the western United States. The topics of discussion include; the Gold Rush; the Comstock lode, NV; Butte, MT; Bunker Hill, Idaho; Leadville, CO; and Bisbee, AZ.

Ripley, Earle A., et al. *Environmental Effects of Mining.* Delray Beach, FL: St. Lucie Press, 1996.
This book surveys environmental impacts of hardrock and coal mining, with a focus on Canadian mining. In the second half of the book, the authors examine production methods, environmental damage, and reclamation techniques at mines extracting and processing commodities such as base and precious metals, coal, diamonds, asbestos, salt, barite, clays, lime, and gypsum. The book uses some case studies to illustrate mining's environmental impacts and reclamation techniques.

Robertson, A.M.. *Long Term Prevention of Acid Mine Drainage.* Proceedings of the International Conference on Control of Environmental Problems from Metal Mines, Roros, Norway, 20-24 June 1988. Oslo, Norway: State Pollution Control Authority, 1988.
This paper describes the physical, chemical, and biological factors which lead to the generation of acid mine drainage (AMD) in waste rock and tailings, as well as how AMD generation differs in these two major forms of mining waste. The paper also discusses methods of preventing AMD and stopping its migration.

U.S. Congress, Office of Technology Assessment. *Copper: Technology and Competitiveness,* OTA-BP-0-82. Washington, D.C., September, 1988.
This report provides an overview of technologies employed in modern copper production. The report describes the conventional sequence of extraction, communition, beneficiation, smelting, and electrorefining steps used to produce copper. The report also describes the more recently developed solvent extraction/electrowinning process for producing copper.

U.S. Congress, Office of Technology Assessment. *Managing Industrial Solid Wastes from Manufacturing, Mining, Oil and Gas Production, and Utility Coal Combustion-Background Paper,* OTA-BP-0-82. Washington, D.C., February, 1992.
This report compares solid waste generation among various industrial sectors, including hardrock mining. The report also presents an overview of the federal and state laws and regulations which apply to solid wastes from each industrial sector. The report's section on hardrock mining describes how most mining wastes are exempted from the strict "cradle to grave" regulations on hazardous wastes of the federal Resource Conservation and Recovery Act (RCRA).

U.S. Department of the Interior. *International Land Reclamation and Mine Drainage Conference and Third International Conference on the Abatement of Acidic Drainage.* Four Volumes. Bureau of Mine Special Publication SP 06A-94.

This collection of papers by technical experts describes various methods of predicting, preventing, controlling, and treating acid mine drainage. Topics covered include capping and submerging of mine wastes, alkaline additions, wetlands treatment, and reclamation.

U.S. Environmental Protection Agency. *Acid Mine Drainage Prediction.* EPA 530-R-94-036. Washington, D.C., December, 1994.

This report presents an overview of current methods used to predict acid mine drainage generation in hardrock mining wastes. The report describes various types of static and kinetic testing — the two major methods of predicting acid mine drainage. The report also presents case studies of mines where the potential to generate acid was not expected or considered.

U.S. Environmental Protection Agency. *Profile of the Metal Mining Industry.* EPA/310-R-95-008. Washington, D.C., September 1995.

This overview of the metal mining industry documents hundreds of unpermitted releases into the environment of acid, cyanide solution, sulfuric acid, and other contaminants at metal mines. This report also discusses waste handling methods that mines can employ to prevent acid mine drainage, cyanide contamination, and other mining-caused pollution.

U.S. Environmental Protection Agency. *Report to Congress, Wastes from the Extraction and Beneficiation of Metallic Ores, Phosphate Rock, Asbestos, Overburden from Uranium Mining, and Oil Shale.* Washington, D.C., 31 December 1985.

This Congressionally-mandated study of the hardrock mining industry describes the massive amounts and types of solid and hazardous waste generated by various mining sectors (metallic and non-metallic), and waste handling techniques commonly used in the industry. In addition, the study reports on EPA finding of ground and surface water contamination with metals, acid, and cyanide at several mine sites.

Warhurst, Alyson. *Environmental Degradation from Mining and Mineral Processing in Developing Countries: Corporate Responses and National Policies.* Paris: Organization For Economic Cooperation and Development, 1994.

This report describes the environmental damage caused by metal mining in developing countries, as well as the political and socio-economic context in which mines operate in these countries. The report also provides some recent examples of the measures which developing countries are taking to better regulate mining's adverse impacts on water, land, and air.

♦ ♦ ♦

AUTHORS

Carlos D. Da Rosa, J.D., is Senior Research Associate at Mineral Policy Center. Da Rosa's prior experience includes work for Congressman Richard Gephardt, the Center for International Environmental Law, the U.S. Department of Justice and the U.S. Environmental Protection Agency. He is a graduate of Yale University and Northeastern University School of Law.

James S. Lyon is Director of Community-Based Programs at the National Wildlife Federation, and former Vice President for Policy at Mineral Policy Center. Lyon has worked with public interest organizations to improve mining regulation for more than fifteen years. He is a graduate of Moravian College, Bethlehem, Pennsylvania.

ACKNOWLEDGMENTS

The authors and Mineral Policy Center wish to thank the multitude of able, dedicated, and generous souls whose support for and participation in the creation of *Golden Dreams, Poisoned Streams* were essential to its completion.

This book has benefited from the cumulative body of research and data base that Mineral Policy Center has assembled since the Center's founding in 1988. Major funding for these general research programs has been provided by the American Conservation Association, Compton Foundation, General Service Foundation, W. Alton Jones Foundation, The Joyce Foundation, Richard King Mellon Foundation, Charles Stewart Mott Foundation, The New-Land Foundation, The Pew Charitable Trusts, Public Welfare Foundation, Rockefeller Family Fund, Town Creek Foundation, Turner Foundation, and the Wilburforce Foundation. Sweet Water Trust provided specific funding for the color photo insert and graphics which enhance this book.

The Educational Foundation of America and the Surdna Foundation have been particularly generous with both their funding support and their personal encouragement to the Center, through thick and thin. Gilman Ordway has been an insightful friend and a committed benefactor. Thank you, friends.

Many Mineral Policy Center staff members, past and present, made irreplaceable personal contributions to this effort. President Phil Hocker provided the vision, guidance, and unswerving commitment to high standards that initiated and molded this book. Vice President for Policy Stephen D'Esposito oversaw its final management and distribution. As Project Coordinators, Stephanie Osborn guided this project through the drafting stage and Dori Gilels managed the completion of the book as well as our efforts to get the book into as many hands as possible.

Interns and other staff at the Center researched and fact-checked the data in this book with persistence and ingenuity: Aimee Boulanger, Susan Brackett, Andrew Carter, Dan Hirschman, Jill Klein, Gary Kravitz, Heather Langford, Kelly Maroti, Tim McCrae, Will Patric, Ed Piasecki, Jennifer Pierce, Halley Rosen, Kris Szatmary, Debby Tipton, Ellen Wertheimer and Rhonda Williams.

Our appreciation is also extended to the variety of experts whose intelligent, carefully-researched papers compose Part II of the book, and to the editing skill and hard work of Andrea Fine, Bob McCoy, and Velma Smith.

Finally, thank you to new and old friends whose technical advice and review of our work helped make *Golden Dreams, Poisoned Streams* accurate and complete: John Burke Burnett, David Chambers, Gail Charnley, Doris Balvín Díaz, T.R. Hathaway, Ann Maest, Jim McElfish, Tom Myers, D. Kirk Nordstrom, and, with special respect, Glenn C. Miller.

MINERAL POLICY CENTER

Mineral Policy Center leads the fight to end environmental damage from mineral development. The Center is a small non-profit organization dedicated to solving the environmental problems caused by the mineral industry, and to preventing their repetition. The group was founded in 1988 by its President, Philip Hocker, together with former Secretary of the Interior Stewart L. Udall. Its headquarters are in Washington, DC; its reach is global.

To win the fight to protect our waters and lands from damage from mineral development, Mineral Policy Center:

- Researches and publicizes the massive environmental problems which mining causes, and the special environmental loopholes and financial subsidies current laws give to mining companies.
- Educates and assists community groups concerned about environmental problems from mining projects.
- Lobbies for reform of the 1872 Mining Law, the Clean Water Act, and waste management laws — to improve the environmental standards required of the mineral-development industry.

Mineral Policy Center is recognized as the leading group advocating environmental reform of mining and mineral extraction in America and around the globe. The Center does not work alone, however; the Center's staff assists many independent local and regional citizens groups concerned with mining projects. Public-interest groups from Ireland to Papua New Guinea and Kyrgyzstan have benefited from the Center's information and guidance on mining issues, as well as organizations in virtually every one of the United States.

Mineral Policy Center's recent successes have included persuading the 105th Congress to sustain a moratorium on the 1872 Mining Law's \$5-per-acre giveaways. In collaboration with other conservation groups, the Center also defeated proposed development of the "New World" gold mine next to Yellowstone National Park. The Center's recommendation that mining be required to report hazardous-materials releases in the EPA's "Toxics Release Inventory" program was made mandatory in 1997.

Mineral Policy Center's work is supported by membership dues, private donations and foundation grants. Dues are \$25/year and are tax-deductible. Please become a member, and join us in this important fight.